Springer Theses

Recognizing Outstanding Ph.D. Research

Aims and Scope

The series "Springer Theses" brings together a selection of the very best Ph.D. theses from around the world and across the physical sciences. Each volume has been selected for its scientific excellence by an internationally renowned institute or department. For greater accessibility to non-specialists, the published versions include an extended introduction, as well as a foreword by the student's supervisor explaining the special relevance of the work for the field. The series as a whole provides a valuable resource both for newcomers to the research fields, and for other scientists seeking detailed background information on special questions. Finally, it serves as an accredited documentation of the valuable contributions made by today's younger generation of scientists.

How to Apply

Theses prepared at internationally leading universities and institutes may be nominated for publication by supervisors with the endorsement of the head of department. Each such department is entitled to nominate one thesis per year. Enquiries and requests for a nomination form should be sent, together with the thesis, to Ms. Nathalie Berg (see link on right).

Eligibility Criteria

To qualify for consideration, theses should fulfil the following criteria:

- They must be written in good English.
- The topic should fall within the confines of Chemistry, Physics, Earth Sciences, Engineering and related interdisciplinary fields such as Materials, Nanoscience, Chemical Engineering, Complex Systems and Biophysics.
- The work reported in the thesis must represent a significant scientific advance.
- A maximum of 30% of the text may be a verbatim reproduction from the author's previous publications.
- They must have been examined and passed during the 12 months prior to nomination.
- The theses should have a clearly defined structure including an introduction accessible to new PhD students and scientists not expert in the relevant field.
- After acceptance, theses should be supplemented by a "Supervisor's Foreword" outlining the wider scientific relevance of the work

Indexed by zbMATH.

Anamaria Hell

The Massless Limit of Massive Gauge Theories

To the Strong Coupling and Beyond

Springer

Anamaria Hell
Kavli Institute for the Physics and
Mathematics of the Universe
The University of Tokyo
Kashiwa, Japan

ISSN 2190-5053 ISSN 2190-5061 (electronic)
Springer Theses
ISBN 978-3-032-01089-6 ISBN 978-3-032-01090-2 (eBook)
https://doi.org/10.1007/978-3-032-01090-2

This Springer imprint is published by the registered company Springer Nature Switzerland AG
The registered company address is: Gewerbestrasse 11, 6330 Cham, Switzerland

For Slava, Dieter, and George.

Foreword

It is with great pleasure that I reflect on the academic journey of Anamaria Hell at the Ludwig-Maximilians-Universität in Munich. Even before the start of her PhD studies, I had the opportunity to interact with her during her time in the Elite Master's program in Theoretical and Mathematical Physics. From the outset, Anamaria demonstrated exceptional ambition and enthusiasm as she entered the field of theoretical physics, particularly in addressing one of its longstanding challenges: the massless limit of massive gauge fields.

Before the discovery of the Higgs mechanism, one of the central questions in high-energy physics was how to consistently endow vector bosons with mass without violating renormalizability and unitarity. Simply adding a mass term by hand—as was done by S. Glashow in his original formulation of the electroweak theory—led, within the standard approach using relativistic gauges, to an apparent discontinuity. Specifically, the perturbative expansion appeared to become singular in the zero-mass limit.

Although there were early attempts to understand this issue—see, for example, the work of Vainshtein and Khriplovich (1971), which attributed the problem to the behavior of the longitudinal mode entering a strongly coupled regime (a phenomenon later recognized as the "Vainshtein scale" and first discovered by Vainshtein in the context of massive gravity in 1972)—there was no systematic analysis of this issue in the literature for general theories exhibiting a discontinuous massless limit.

Anamaria developed a novel approach to this problem. Rather than working within the explicitly Lorentz-invariant formalism, she employed a reduced symmetry group—namely, the rotation group—to isolate the problematic longitudinal mode, adopting techniques from cosmological perturbation theory. This allowed her to identify the strong coupling scale at which the longitudinal mode enters the non-perturbative regime and to demonstrate that, at this scale, the mode effectively decouples from the remaining sector. She further quantified the corrections due to this mode to the dynamics of the remaining perturbative degrees of freedom.

Through this approach, she was able to show that the massless limit can in fact be smooth in theories previously thought to possess a van Dam–Veltman–Zakharov-type discontinuity. Anamaria examined several such theories, presenting her results in this volume with exceptional pedagogical clarity.

After reading her paper on the electroweak theory—originally introduced by Glashow in 1961 with mass terms added by hand—Professor Glashow wrote to her:

> I find your result to be most surprising and fascinating. Evidently, I was not so stupid as I had thought.

In summary, Anamaria's thesis introduces both innovative techniques and profound conceptual insights into quantum field theory. Her contributions challenge established assumptions and provide a fresh perspective on foundational problems. These ideas are already influencing ongoing research and will undoubtedly remain relevant to the future development of theoretical physics.

Munich, Germany
June 2025

Viatcheslav Mukhanov

Preface

This thesis studies the massive non-linear gauge theories with mass added *by hand*. First, we will consider the massive Yang-Mills theory, the theory of a non-Abelian vector field. The conventional approaches suggest that this theory does not have a smooth massless limit. To gain further insight into this theory, we will also consider a toy model of Proca theory with a cubic self-interaction, and show that this theory also suffers from an apparent perturbative discontinuity. By using the decomposition of the vector field into the scalar and vector modes, based on the rotation group, we will confirm that the previous issues arise due to the longitudinal modes. In contrast to the traverse modes, these do not appear in the massless theory. However, by taking into account the non-linear terms, we will show that due to them, the longitudinal modes become strongly coupled, meaning that the perturbation theory breaks down. This takes place at the Vainshtein scale, which coincides with that of unitarity violation. Beyond it, we will see that the longitudinal modes remain strongly coupled and decouple from the remaining degrees of freedom up to small corrections. Thus, we will find that the massless limit of the massive Yang-Mills theory is smooth, as conjectured by A. I. Vainshtein and I. B. Khriplovich.

We will then extend our study to the theory of a massive antisymmetric two-form—the Kalb-Ramond theory—and a theory of a massive three-form. We will compare the two theories with Proca theory and a theory of a massive scalar field, respectively, modifying them by a quartic self-interaction. As a result, we will find that all theories apart from the massive scalar theory have the same Vainshtein scale. However, as we will see, the degrees of freedom of the two pairs of theories do not behave the same. In Proca theory, the longitudinal mode enters a strong coupling regime and decouples from the transverse modes, which remain weakly coupled and survive in the massless limit. In contrast, in Kalb-Ramond theory, the transverse modes become strongly coupled and decouple from the longitudinal mode that is in the weak coupling regime and survives in the massless limit. Similarly, we will show that the pseudoscalar—the degree of freedom of a massive three-form—becomes strongly coupled. Thus in the massless limit, this theory has no propagating degrees of freedom. These results indicate a contradiction with numerous claims in the literature, that the theories of massive Kalb-Ramond and Proca field and theories of massive three-form and a scalar field are dual.

Kashiwanoha, Japan
June 2025

Anamaria Hell

Acknowledgments

The words *thank you* are not enough to express my gratitude to all my mentors, friends, and family, who have so far supported me, guided me, and taught me, on the way to achieving my goals.

First and foremost, I would like to express my deepest gratitude and admiration to Prof. Viatcheslav Mukhanov, my PhD supervisor, for his continued support—even beyond my stay in Munich—as well as for his inspiration, encouragement, guidance, and patience, and for giving me the opportunity to learn. I will happily cherish all of the discussions, amazing lectures and stories, and his insightful perspective on physics.

I would also like to especially thank Prof. Dieter Lüst, for amazing collaboration, his support, warmth, and guidance, and for inspiring me to push forward, in both research and organization of schools and workshops.

In addition, a special *thank you* goes to Prof. George Zoupanos, who has without a doubt played an important role in my education as well, through wonderful discussions on both life and physics.

I believe that the interactions with Robert H. Brandenberger, Gerhard Buchalla, Ali Chamseddine, Gia Dvali, Michael Haack, Stefan Hofmann, Anna Ijjas, Ivo Sachs, Paul J. Steinhardt, Arkady Vainshtein, and Alex Vikman were particularly important for my work and growth during my PhD and afterward. Therefore, I would like to sincerely thank them for this.

I would also like to thank my colleagues, and friends, Ivana Babic, Cecilia Giavonni, Katrin Hammer, Tobias Russ, Thomas Steingasser, Jan Toelstede, Pablo Sala, and the Baier family for everything.

This book has been written several years after the original PhD thesis. Therefore, I would also like to use this chance to especially thank Misao Sasaki, Elisa Ferreira, Saeko Hayashi, Yukari Ito, and Tadayuki Takahashi for their support in my academic life, great discussions, and opportunities to learn and grow. Specially, I would also like to thank Betti Hartmann, for encouraging me and supporting me to write this book.

Finally, I would like to express my thanks to my family, and especially my mother, Kristina, for always being there for me.

Competing Interests The author has no competing interests to declare that are relevant to the content of this manuscript.

Contents

Acronyms

SM	Standard model of particle physics
GR	General relativity
mYM	Massive Yang-Mills
YM	Yang-Mills
dof	Degrees of freedom

Chapter 1
Introduction

One of the most exciting directions in modern physics is to consider modifications to the well-known theories that are, to date, confirmed by observations in the leading order approximation. The first such theory is the Standard Model of Particle Physics (SM). It describes fundamental particles and excitations of the fields that are based on the gauge group $SU(3) \times SU(2) \times U(1)$. This theory studies the main building blocks of matter, from electrons moving around the atomic nucleus to the quarks that form protons and neutrons, which then make the atom's core. It also accounts for three forces in nature—the weak, strong, and electromagnetic forces. The fourth force, gravity, is described by General Relativity (GR), a theory of the dynamics of space-time. It studies how matter affects the curvature of spacetime, and in turn, how gravity influences matter.

Both SM and GR serve as a cornerstone for theoretical physics. Yet, we know that the two are not the final theories of nature—our understanding of the Universe is still challenged by several puzzles. One of the major ones is to understand what is the substance that drives its acceleration. If considered at early times, then the theory that might account for it is known as inflation [1–7]. This theory is particularly appealing, as it can give rise to the seeds for the structure of the Universe that we see today, and at the same time agrees with the Cosmic-Background-Radiation (CMB) measurements [8–14]. In addition to the rapid expansion at the very first moments, observations suggest that the Universe accelerates also at late times [15, 16]. The substance that drives it is known as dark energy. It can be either a cosmological constant, an explicit constant in the action of Einstein's gravity, or a dynamical fluid, whose values change over time [17–28]. Moreover, the matter content found in the Universe presents another mystery. Mesurements suggest that it contains a dark substance, known as dark matter, which accounts for its missing mass [29–31].

To date, many models aim to underpin the theories of dark energy, dark matter, and inflation, among many other puzzles in high-energy physics. Most of these are based on modifications to the standard model theories, by extending the particle

A. Hell, *The Massless Limit of Massive Gauge Theories*, Springer Theses,
https://doi.org/10.1007/978-3-032-01090-2_1

physics sectors, or by considering modified theories of gravity, which reduce to GR only in certain limits. Sometimes, these modifications may appear to be trivial. However, at the same, they could give rise to significant and unexpected consequences. In this thesis, we will study one of the simplest, and historically important examples of such a case—altering the theory by adding the mass to the fields.

1.1 Degrees of Freedom and Mass

When studying a theory of massless fields, one of the most natural questions is to ask how the theory changes if the fields are massive. This is also tempting, as the observations in principle allow them to be massive—they provide only an upper bound on this value. This question might be trivial, however, if one considers scalar fields. For example, a theory of a massless scalar field $\phi(\boldsymbol{x}, t)$ is described by the following Lagrangian density:

$$\mathcal{L} = \frac{1}{2}\phi_{,\mu}\phi^{,\mu}, \tag{1.1}$$

where $_{,\mu} = \partial_\mu = \frac{\partial}{\partial x^\mu}$ is the partial derivative with respect to the coordinate x^μ. To make it massive, we simply introduce a mass term *"by hand"* to the Lagrangian:

$$\mathcal{L} = \frac{1}{2}\phi_{,\mu}\phi^{,\mu} - \frac{1}{2}m^2\phi^2. \tag{1.2}$$

We can see that the resulting theory still describes a single scalar field, which now however propagates with a speed smaller than that of light due to its mass. If we take it to zero we obtain the massless theory with ease. While the question on massive fields seems trivial on this level, we will see that it becomes much more interesting as soon as we consider vector or tensor fields.

The core of this lies in the number of degrees of freedom—the number of initial conditions divided by two [32]. By studying the equations of motion for the above scalar field, which is given by

$$\left(\Box + m^2\right)\phi(\boldsymbol{x}, t) = 0, \tag{1.3}$$

where $\Box = \partial_0^2 - \Delta$ is the wave operator, and $\Delta = \partial_i\partial_i$ is the Laplacian, we can notice that we need to determine two initial conditions for the scalar field, per point in space-time—its initial position and its initial velocity. Thus, the above scalar field has only one degree of freedom, per point in space-time.

The study of the degrees of freedom becomes slightly more complicated if one considers a vector or a tensor field. In particular, electrodynamics has two degrees of freedom, which correspond to the two polarizations of photons. At the same time, to describe it, one uses a vector field $A_\mu(x,t)$, which has four components, instead of two, due to the space-time index μ that takes values from 0 to 3. A quick way to see the degrees of freedom is to study the Lagrangian density,[1] similarly to [33]. For the case of electrodynamics, the action is given by:

$$S = \int d^4x \left(-\frac{1}{4} F_{\mu\nu} F^{\mu\nu} \right), \tag{1.4}$$

where

$$F_{\mu\nu} = A_{\nu,\mu} - A_{\mu,\nu} \tag{1.5}$$

is the field-strength tensor. To understand which components of the vector field are important for the dynamics, let us separate the temporal and spatial components:

$$(\ A_0,\ A_i\ ), \qquad \text{with} \qquad i = 1, 2, 3, \tag{1.6}$$

and, further decompose the spatial part in the following way:

$$A_i = A_i^T + \chi_{,i}, \qquad \text{where} \qquad A^T_{i,i} = 0. \tag{1.7}$$

Then, we find the following Lagrangian density:

$$\mathcal{L} = \frac{1}{2}\left(-A_0 \Delta A_0 + 2A_0 \Delta \dot{\chi} - \dot{\chi} \Delta \dot{\chi} + \dot{A}_i^T \dot{A}_i^T - A^T_{i,j} A^T_{i,j} \right). \tag{1.8}$$

We can notice that among all of the components, the temporal one is special—it does not have any kinetic term with the form $\dot{A}_0 \dot{A}_0$. This means that it is constrained, depending on other fields. By varying the action with respect to it, we find that it satisfies the following constraint:

$$\Delta A_0 = \Delta \dot{\chi}. \tag{1.9}$$

By solving it, and substituting it back to the action, we find the following expression:

$$\mathcal{L} = \frac{1}{2}\left(\dot{A}_i^T \dot{A}_i^T - A^T_{i,j} A^T_{i,j} \right), \tag{1.10}$$

which indicates only two degrees of freedom, as the transverse modes A_i^T are divergence-less. The remaining longitudinal mode has disappeared from the above expression, as it got canceled by the contribution of the temporal component.

[1] One can check that the procedure is equivalent to the Hamiltonian formalism.

Curiously, however, if the vector field acquires mass by hand, the number of degrees of freedom changes. The theory that describes this massive vector field is known as the Proca theory [34], with action given by:

$$S = \int d^4x \left(-\frac{1}{4} F_{\mu\nu} F^{\mu\nu} + \frac{m^2}{2} A_\mu A^\mu \right). \tag{1.11}$$

As we will see in the following chapter, in this case, the constraint for the temporal component becomes modified, such that the scalar χ, also known as the longitudinal mode, is no longer cancelled. Rather, it propagates in addition to the two transverse modes. The addition of degrees of freedom in massive theories when compared to the massless ones is not only connected to vector fields. Rather, as we will study in this thesis, it also appears for more general tensor fields and even p-forms.

One can also wonder if there is a way to make theories massive, by at the same time preserving their number of degrees of freedom. This is known as the Brout-Englert-Higgs mechanism [35–37]. However, in contrast to the previous case, it requires us to initially start with more fields than the one that we aim to make massive. For example, in the case of an Abelian vector field, we can consider the following coupling case:

$$\mathcal{L} = -\frac{1}{4} F_{\mu\nu} F^{\mu\nu} + \frac{1}{2} D_\mu \phi \left(D^\mu \phi \right)^* - V(\phi\phi^*), \tag{1.12}$$

where

$$D_\mu = \partial_\mu + i\lambda A_\mu \tag{1.13}$$

is the covariant derivative, and ϕ is a complex scalar field. We can notice that the vector field in this case interacts with the scalar, through the following interaction:

$$\lambda^2 \phi\phi^* A_\mu A^\mu. \tag{1.14}$$

Thus, if the scalar field potential has a minimum for some value, we can expand the scalar field around it, and automatically generate a mass term for the vector field.[2] In this way, the number of degrees of freedom remains preserved—one starts initially with four modes, with two corresponding to the scalar field, and remains with four, since the vector field becomes massive, and only the radial part of the scalar field remains in the theory.

The Higgs mechanism has revolutionized the study of weak interactions. It allowed the gauge bosons, the non-Abelian gauge fields that are based on the SU(2) group, the become massive in such a way that the resulting quantum field

[2] One of the best ways to see this is with an introduction of gauge-invariant fields with respect to the U(1) transformations, as shown in [38], for both Abelian and non-Abelian theory.

theory satisfies two key properties—unitarity and renormalizability, which were satisfied also in the massless Yang-Mills theory [39, 40]. The first property secures the conservation of probability when studying scattering processes, such that the total probability of all possible outcomes sums to unity. The second deals with divergences that arise when studying quantum corrections, that typically appear in loop diagrams. In particular, if such divergences can be absorbed by a finite amount of parameters, we say that the theory is renormalizable [41, 42].

In order to describe the weak interactions, one needed the SU(2) Yang-Mills fields to be massive, while at the same time satisfying the two properties. However, this did not seem to be easy to realize if one simply added a mass term by hand, as it was initially proposed [43]. Instead, by coupling the theory to the complex scalar fields, one can make the gauge fields massive, and coupled to the Higgs field, while the theory at the same time preserves the two properties, making it ultravioletly complete.

One can nevertheless wonder, at least as a theoretical puzzle, or as a possible realization in the dark sector, what happens if one simply adds the mass term by hand. One of the goals of this thesis is to explore this question.

1.2 Small Changes and Big Consequences

According to the principle of continuity, introduced by L. Bass and E. Schrödinger in 1955., if one alters a physical theory, then regardless of the size of this modification, all observables should smoothly approach the original ones once one takes the limit back to the initial theory [44]. Yet, massive Yang-Mills theory (mYM), with mass added *"by hand"*, has seemed to defy this principle. Despite its simplicity, when studying such a theory one soon encounters a series of difficulties for a small mass.

Similarly to the Abelian case, if one modifies the massless theory by an explicit mass term, one finds additional degrees of freedom per group color—scalars, known as the longitudinal modes. In principle, that should not necessarily indicate a problem with the theory. However, the propagator of this theory:

$$\tilde{\Delta}^{ab}_{\mu\nu}(k) = \left(-\eta_{\mu\nu} + \frac{k_\mu k_\nu}{m^2}\right) \frac{i\delta^{ab}}{k^2 - m^2 + i\epsilon}, \tag{1.15}$$

a two-point correlation function for the gauge fields becomes constant at high energies, thus indicating a power-counting non-renormalizable theory. Moreover, the above propagator is clearly singular in mass, in contrast to the massive scalar field, whose propagator is simply given by:

$$\tilde{\Delta}_\phi(k) = \frac{i}{k^2 - m^2 + i\epsilon}. \tag{1.16}$$

Thus, one easily encounters behavior that is singular in mass within the perturbative approach. As we will see in detail in the following chapter, the reason why this term with inverse mass behavior appears lies in the longitudinal modes. Curiously, the Proca theory has the same propagator. However, in contrast to the previous case, this theory is not self-interacting. Thus, if coupled to the conserved source, it can be shown that it is renormalizable [45]. Due to the presence of self-interactions, that arise when studying the non-Abelian gauge fields, the mYM theory, on the other hand, can be shown to be non-renormalizable already when studying the two-loop corrections [46–53].

Moreover, by applying standard perturbative methods, one can find that the unitarity is violated at energy scales [50, 54–58]:

$$k_u \sim \frac{m}{g}, \tag{1.17}$$

where g is the coupling constant of the theory.

Notably, by studying quantum correction to the propagator of the vector field, in [53] it was noticed that for two or more loop diagrams, the singularities in mass form a series

$$\frac{g^2 \Lambda^2}{m^2}, \tag{1.18}$$

with Λ being the cut-off scale. Thus, it was proposed that the series could be re-summed, such that all singularities disappear after the ressumation, curing the apparent discontinuity in the mass. However, the next year, the authors of [59] took back this possibility. In particular, by studying the imaginary parts of the one-loop corrections to the transverse modes which are finite in contrast to the higher order corrections, they have noticed that upon taking the massless limit, the corrections differ from the massless case by a factor of two:

$$\text{Im}\,(D_2) = 2\ \text{Im}\,(D_1)|_{m \to 0}\,, \tag{1.19}$$

where D_1 and D_2 denote the following diagrams:

Thus, given this finite discontinuity, they concluded that the massless limit of the massive Yang-Mills theory is not smooth. This is due to the longitudinal modes, that fail to decouple from the remaining degrees of freedom. Therefore, even if one decides to modify the massless YM theory slightly, by adding an almost vanishing mass term, it would appear that the main properties of the massless theory—its unitarity and renormalizability are lost, according to the conventional methods, and that one cannot obtain its massless version, by simply taking $m \to 0$ (Fig. 1.1).

A similar effect, and an even more drastic one, was found also in linearized massive gravity—a theory of a massive spin-2 particle with the mass term of the Fiersz-Pauli form [60]. Its predictions were shown to deviate from those of General Relativity to such an extent that the theory could be ruled out as a possible theory

Fig. 1.1 The 1-loop corrections to the propagator for the vector modes. The D_1 corresponds to the massless case, and the D_2 to the massive one

of nature. In particular, as shown in [59, 61, 62], if one evaluates the massive linearized gravity up to leading order corrections, one can find that the precession of perihelia of Mercury or the deflection of starlight differ from linearized gravity by factors that survive in the massless limit. This is known as the van Dam-Veltman-Zakharov (vDVZ) discontinuity. Similarly to the mYM case, when modifying the linearized gravity, which describes two tensor modes, with a mass term, new degrees of freedom arise—a scalar degree of freedom and two vector ones, making the linearized massive gravity a theory of five degrees of freedom. The previous discrepancy emerges due to the longitudinal mode, a degree of freedom of massive gravity, absent in the massless theory, which fails to decouple from the remaining degrees of freedom once the mass is taken to zero.

1.3 The Non-linear Terms

The previous findings are strange—they imply that the mass of the graviton should be strictly equal to zero, or that one should not explicitly introduce the mass for the Yang-Mills fields explicitly in the Lagrangian. Yet, in accordance with the principle of continuity, one would not expect that adding a mass term as small as $m = 10^{-32}$ eV should make an important difference. Therefore, it is important to question if the previous results always hold, or if we are missing something instead.

When studying fields in nature, we almost never encounter free fields. As an example, for the scalar field, that means that it is possible to add also a non-trivial potential to the Lagrangian density:

$$\mathcal{L} = \frac{1}{2}\phi_{,\mu}\phi^{,\mu} - \frac{1}{2}m^2\phi^2 - \frac{1}{4}\lambda^2\phi^4, \tag{1.20}$$

where λ is the coupling constant. In order to study the above theory, one cannot directly evaluate the non-linear term. Rather, one assumes that its contribution towards an overall dynamics is much smaller than the two quadratic terms. Equivalently, this corresponds to the linearized terms on the level of the equations of motion:

$$\left(\Box + m^2\right)\phi = \lambda\phi^3 \tag{1.21}$$

Essentially, given this assumption, one can then apply the perturbation theory. For this, one can assume that the scalar field can be expanded as

$$\phi = \phi^{(0)} + \phi^{(1)} + \dots \tag{1.22}$$

where $\phi^{(0)}$ satisfies the linearized equation:

$$\left(\Box + m^2\right) \phi^{(0)} = 0. \tag{1.23}$$

Given this, we can then calculate the first order corrections to the previous result, given by $\phi^{(1)}$, which satisfies:

$$\left(\Box + m^2\right) \phi^{(1)} = \lambda^2 \left(\phi^{(0)}\right)^3, \tag{1.24}$$

or the second-order ones. However, these results are trustable only if the previous assumption holds—the linear term $\phi^{(0)}$ should be much more dominant then the corrections. If the opposite holds, it would mean that our perturbation theory is not valid, as it would be opposed to the initial assumption. Notably, while at high energies, the above perturbation theory holds as long as the coupling of the scalar λ is much smaller than unity, other theories might give rise to more complicated conditions, that also depend on the length scales, energy scales and even the mass of the field.

Similarly to the previous case, both GR and YM theory are non-linear theories. Moreover, this holds the corresponding massive theories as well. Therefore, there could be a scale, known as the strong coupling scale, at which the perturbative approach to the previous theories breaks down. Notably, as pointed out in [63], in the case of massive graivity, this takes place at the Vainshtein radius—a scale at which the non-linear terms become of the same order as the linear ones. One this scale is reached, the longitudinal mode that gives rise to the previous perturbative discontinuity becomes strongly coupled, looses its linear propagator, and decouples from the remaining degrees of freedom. Beyond the Vainshtein radius, one recovers the predictions of GR up to very small scales, with the scale stretching to infinity as the mass of the graviton tends to zero. This mechanism is known as the Vainshtein mechanism, introduced very soon after the introduction of the vDVZ discontinuity [63–65].

Despite its success in massive gravity, until recently, the puzzle of massive Yang-Mills theory has still remained. In contrast to the case of massive gravity, in this theory apparent discontinuity in the massless limit appears for one-loop corrections. Notably, A. Vainshtein and I. B. Khriplovich have conjectured that it might be cured beyond the perturbation theory [58]. Yet, until recently, the proof of this was still unknown. One of the main aims of this thesis is to show that the massless limit of this theory is smooth, in spite of the conclusions drawn from the conventional approaches (Fig. 1.2).

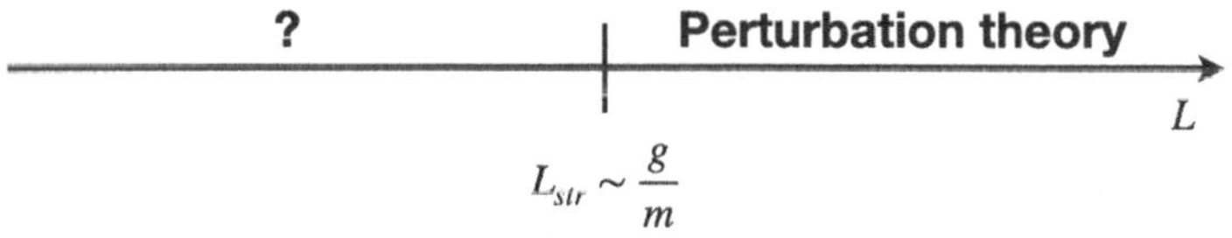

Fig. 1.2 Schematic representation of the strong coupling and the validity of the perturbation theory

By utilising the methods of cosmological perturbation theory, we will study the underlying structure of massive Yang-Mills theory following the work done in [66]. We will confirm that the source of the discontinuity are the longitudinal modes, the degrees of freedom that are not present in the massless theory. Moreover, we will find that the perturbation theory for these modes is not valid up to all possible energies. Rather, once the non-linear terms become of the same order as the linear ones, at the Vainshtein scale, they will become strongly coupled. Therefore, we will find the following picture:

The perturbation theory will only be valid up to the strong coupling length scale, L_{str} which depends on the mass and the coupling of the theory. Notably, this scale will match the scale at which the unitarity is violated. In the sense of effective theories, one can study the theory only up to this scale. However, we will show that this theory has an additional feature—one can go even beyond this scale, and investigate the effects of the degrees of freedom. In this case, we will find that the longitudinal modes lose their linear propagator, and become strongly coupled. Moreover, we will find that the transverse modes in turn remain weakly coupled for scales above and beyond the Vainshtein scale. The corrections that appear due to the longitudinal modes will become smaller as we approach higher energies. Ultimately, we will see that the corrections dissapear in the massless limit, with the strong-coupling length scale becoming infinite. As a result, we will find that the discontinuity in the massless limit was nothing more than an artefact of the perturbation theory, as initially proposed in [58].

1.4 The Origin of the Strong Coupling

As we will find in the upcoming chapters, the Vainshtein mechanism shows an underlying connection between massive Yang-Mills theory and massive gravity—both of these theories have an apparent perturbative discontinuity which appears due to the longitudinal modes—degrees of freedom that are absent in the corresponding massless theories. However, due to their non-linear nature, such modes become strongly coupled and decouple from the remaining modes, for which in turn, the perturbative approach holds, thus recovering the massless theories up to small corrections. Notably, in addition to the longitudinal modes, in the mimetic massive gravity—an alternative theory of a massive graviton, it was shown that the vector modes become strongly coupled as well [67].

This suggests a trend—if one modifies a theory, and as a result introduces additional degrees of freedom, then, if one encounters a perturbative series that is singular in the parameter characterizing these modes (eg. mass), it is very likely that such modes will become strongly coupled.

The origin of this behavior can be traced to the quantum fluctuations of the fields. To understand this, let's go back to the example of a free scalar field, following [42]. If the field is classical, $\phi(\boldsymbol{x}, t) = 0$ is the solution of the equation of motion. However, the same is not true for the quantum theory. In order to find it, we introduce a conjugated momenta associated with the scalar field:

$$\pi = \frac{\delta L}{\delta(\dot{\phi})}, \tag{1.25}$$

given in terms of the functional derivative of a Lagrangian for the scalar field:

$$L = \int d^3x \mathcal{L} = \int d^3x \frac{1}{2}\left(\dot{\phi}^2 - \phi_{,i}\phi_{,i} - m^2\phi^2\right). \tag{1.26}$$

To quantize the theory, one then promotes the scalar field and its conjugated momentum

$$\pi = \dot{\phi} \tag{1.27}$$

into operators:

$$\phi(\boldsymbol{x}, t) \rightarrow \hat{\phi}(\boldsymbol{x}, t) \quad \text{and} \qquad \pi(\boldsymbol{x}, t) \rightarrow \hat{\phi}(\boldsymbol{x}, t), \tag{1.28}$$

and postulate the equal time canonical commutation relations among the two:

$$\left[\hat{\phi}(\boldsymbol{x}, t), \hat{\pi}(\boldsymbol{y}, t)\right] = i\delta(\boldsymbol{x} - \boldsymbol{y}), \tag{1.29}$$

where $\delta(x - y)$ is the Dirac delta function, as well as

$$\left[\hat{\phi}(\boldsymbol{x}, t), \hat{\phi}(\boldsymbol{y}, t)\right] = 0 \qquad \text{and} \qquad \left[\hat{\pi}(\boldsymbol{x}, t), \hat{\pi}(\boldsymbol{y}, t)\right] = 0. \tag{1.30}$$

If we set $\hat{\phi} = 0$, clearly, the above commutation relations will be violated. Rather, in the ground state, the state with the minimal energy, the field fluctuates, with a minimal amplitude of quantum fluctuations:

$$\delta\phi_L \sim \frac{1}{L} \tag{1.31}$$

for scales $L = |\boldsymbol{x} - \boldsymbol{y}| \sim \frac{1}{k}$ evaluated at very high energies $k^2 \gg m^2$.

In contrast to the previous case, the minimal amplitude of quantum fluctuations for modes that appear in massive gauge theories is additionally weighted by the mass. In particular, as we will see in the upcoming chapters, the transverse modes of mYM theory, A_i^T, which are also present in the massless counterpart, fluctuate with the minimal amplitude of the quantum fluctuations that are given by:

$$\delta A_{iL}^T \sim \frac{1}{L}. \tag{1.32}$$

On the other hand, for the longitudinal modes, we will find:

$$\delta \chi_L \sim \frac{1}{mL}. \tag{1.33}$$

Thus, within the perturbative approach, if the field is self-interacting, it can easily give rise to the singular-in-mass behavior. This can take place even for more generalized theories, either modified by a mass or also by additional interactions. By studying such cases, we can uncover new connections among various theories, and also gain new insights that challenge previously thought relations among others.

1.5 Dualities

The Vainshtein mechanism offers a solution to the puzzle of the massless limit in both massive gravity and massive Yang-Mills theory. Nevertheless, the application of this kind of mechanism is not limited only to those theories. By understanding how it manifests within various other massive gauge theories, we will gain a new perspective on the concept that connects some of these theories, known as the duality.

Dual theories are theories, that describe the same physical effect [68]. This notion has recently flourished in the quantum field theory, string theory and the condensed matter physics [69–77]. It has also been extended to the study of massive and massless gauge theories, relating different theories of p-forms.

In particular, in d-dimensions, it is said that a massless p-form theory is dual to a d-p-2 form. Thus, for example, a vector field, which corresponds to a 1-form, is dual to itself in four dimensions, or, in other words, selfdual. The Kalb-Ramond field, an antisymmetric two-form $B_{\mu\nu}$, which is an integral part of the string theory, on the other hand, is said to be dual to a 0-form, or, in other words, to a scalar field. This is closely connected also with the number of degrees of freedom—the massless Kalb-Ramond field describes a single pseudoscalar mode, and a duality transformation, which connects the two theories.

Curiously, while the vector field and the Kalb-Ramond are intristically different fields, if the two theories are massive, they describe same number of degrees of freedom—a longitudinal mode and two transverse ones. As a result, there were numerous claims in the literature about the two theories being dual, and thus describing the same physics [78–96]. Yet, such connection is curious from the point of view of the massless limit. The duality of the corresponding massless theories does not hold as the two theories have different number of the degrees of freedom. Therefore, one could expect that upon modifying them by a mass term, the fields that appear will be weighted down by this parameter, and thus give rise to different physical behavior, traced back to their corresponding amplitude of quantum fluctuations.

In this thesis, based on the study of the massless limit in the mYM theory, we will challenge the concept of dualities, and extend it also to the three-form theories. Following the work done in [97], we will modify the Proca and Kalb-Ramond theories by adding a quartic self-interaction and compare their structure. Thus, as an application of the tools that we will use to show that the massless limit of the massive Yang-Mills theory is smooth, we will find that the two theories exhibit a significantly different behavior, indicating that the original theories might not be dual.

In particular, we will show that in the presence of the self-interactions, within the perturbation theory, the longitudinal mode in the vector field indicates singular behavior of in the mass, and becomes strongly coupled, whereas the transverse modes remain weakly coupled. For the Kalb-Ramond field, on the other hand, we will show that the transverse modes, the degrees of freedom that are absent in the massless theory become in turn strongly coupled, and decouple from the longitudinal modes, which remain in the weakly coupled regime. This analysis can also be generalized for other self-interactions and forms. In addition, we will also investigate the duality of a 3-form and a scalar field using a similar approach.

Thus, while on the one hand, we will uncover connections between mYM theory and massive gravity due to the longitudinal modes, on the other, the notion of strong coupling will lead us towards new findings in relations among different massive and self-interacting gauge theories with respect to the dualities.

1.6 Outlook

This thesis is structured as follows. First, we will focus on the perturbative aspects of the massive Yang-Mills theory. To increase our intuition, in addition to the massive Yang-Mills theory, we will also consider a toy model of self-interacting Proca theory with a cubic self-interaction. Initially, we will study its properties with standard perturbative methods such as unitarity tests. This will allow us to easily generalize the procedures also to Yang-Mills theory, whose properties we will describe in detail at the end of the chapter.

Following the analysis of the perturbative aspects of the massive Yang-Mills theory, we will review the key points of massive gravity and methods that allow us to infer the Vainshtein scale. We will apply them first to a toy model, followed by the massive Yang-Mills theory, where we will resolve the puzzle of its massless limit.

After studying one of the main parts of this thesis—the massless limit of massive Yang-Mills theory, we will focus on relations among different gauge theories, that are known as dualities. In particular, we will investigate if such a notion can be applied to self-interacting theories of p-forms, with special attention to Proca, Kalb-Ramond, and 3-form theories.

The massless limit of massive Yang-Mills theory and its implication on dual theories—the cores of this thesis—are questions that require us to go beyond the standard methods of diagrams and force us to unravel the underlying structure of massive gauge theories. Before we start tackling these questions, let us begin with the basics of conventional methods in quantum field theory and the troubles that arise when one considers a massive vector theory with mass added *by hand*.

Chapter 2
The Perturbative Aspects of Massive Gauge Theories

Yang-Mills theory is a theory of massless non-Abelian vector fields, that has become a central focus of research only a few years after its discovery [98]. It is an $SU(N)$ generalisation of electrodynamics, which in contrast also includes self-interactions of vector fields. However, the first attempts to apply it to physical processes such as the beta decay have soon reached a challenge [99].

The 4-Fermi theory of weak interactions, a theory that aimed to explain this process was only applicable at low energies, up to 300 GeV [100]. The reason for this were quantum corrections. Beyond the leading order, the theory was plagued by infinities that were impossible to eliminate [99], making the theory non-renormalizable. As a result, one had to seek for another theory, that could extend to higher energies and ultravioletly complete the original one. Yang-Mills theory was considered as a possible candidate. However, the vector fields had to be massive, and adding this mass has shown to be more complicated than one could even expect.

The theory of massless Yang-Mills fields is renormalizable and unitary [39]. The first property means that one can absorb all infinities that arise from the loop corrections by a finite amount of terms. The unitarity, on the other hand, reflects the conservation of probability when studying the scattering processes. However, massive Yang-Mills theory, with mass added *by hand*—initially included in the basis of electroweak interactions in [43]—did not seem to satisfy any of these properties. The standard perturbative methods suggested that the theory is neither renormalizable nor unitary [51, 53, 54]. Furthermore, in the massless limit, there was no agreement with the corresponding massless version—the two theories provided different results even for leading corrections, that were regular in the massless limit [59].

Given that such a theory was not a promising step alongside the 4-Fermi theory, the theory of weak interactions was completed only later, after the introduction of the Higgs mechanism [35–37]. This mechanism extended the theory by considering an $SU(2) \times U(1)$ gauge theory, which included also complex scalar field with a potential. Thus, in contrast to simply adding the mass by hand, it conserved the

A. Hell, *The Massless Limit of Massive Gauge Theories*, Springer Theses,
https://doi.org/10.1007/978-3-032-01090-2_2

number of degrees of freedom, before and after the radial part of the scalar acquired the vacuum expectation value, making the vector bosons massive. In particular, initially, one had the massless non-Abelian fields coupled to scalar degrees of freedom, and then ending up with massive vector fields and an additional field, called the Higgs field. Notably, in this way both unitarity and renormalizability was preserved [40]. Yet, while this theory became the main part of the Standard Model of particle physics a puzzle of massive Yang-Mills theory nevertheless remained—*Why adding a mass by hand changes the properties of the theory so drastically?*

In this chapter, we will begin to explore this question. We will investigate the perturbative properties of the massive Yang-Mills theory through conventional methods, that rely on the diagrammatic techniques. In order to simplify the approach, as a first step, we will consider a simpler toy model—a theory of a massive Abelian vector field with cubic self-interaction. This theory will help us to develop our intuition and illuminate the key properties of massive vector fields, while at the same time allowing for an easy generalization to non-Abelian theory. First, we will explore the basics of this theory, including the quantization of this theory. Then, by using the diagramming techniques, we will learn about the subtleties that emerge in these theories, and provide alternatives for approaching them, with the goal of answering the following questions: *Are these theories unitary and renormalizable, and if not—why?*

2.1 The Theories of Massive Vector Fields

Let us now introduce the basic properties of the theories we will examine in the following chapters: a toy model of Proca theory with cubic self-interaction, and the massive Yang-Mills theory, with mass explicitly introduced by and in the Lagrangian.

2.1.1 Free Proca Theory

As a first step, we will consider the free Proca theory [34]. This is a theory of a massive, Abelian vector field, potentially describing a massive photon. One could be puzzled, as it is often thought that the mass of the photon should be strictly zero. However, this is apriori not confirmed, and the photon might have a small non-vanishing mass. Experimentally, we only know its upper bound [101]. Proca theory—the theory of a massive photon—gives us the opportunity to analyze this possibility. This theory is well-known in the standard literature (see eg. [102]). Here, we will review its structure, quantize it and use it to highlight the main properties that will also appear in another theory of massive vector fields—massive Yang-Mills theory.

The action describing a massive photon is given by

$$S = \int d^4x \left(-\frac{1}{4} F_{\mu\nu} F^{\mu\nu} + \frac{m^2}{2} A_\mu A^\mu \right), \tag{2.1}$$

where m is the mass of the Abelian vector field A_μ, and

$$F_{\mu\nu} = A_{\mu,\nu} - A_{\nu,\mu} \tag{2.2}$$

is the field strength tensor. Throughout this work, we will denote the derivatives $\frac{\partial}{\partial x^\mu}$ by a comma $_{,\mu}$. In contrast to Maxwell theory, whose action is given only by the first term of (2.1), Proca theory no gauge redundancy

$$A_\mu \rightarrow \tilde{A}_\mu + \lambda_{,\mu} \tag{2.3}$$

where $\lambda = \lambda(x)$ is a space-time dependent parameter associated with the U(1) gauge group, due to the presence of a mass term. As a result, it is natural to expect that the theory propagates larger number of degrees of freedom. In order to find them, we will use the Hamiltonian formalism for the constrained fields.

2.1.1.1 The Hamiltonian and the Constraints

The Hamiltonian of the theory can be found by using the procedure described in [103]. As a first step, we have to find the conjugated momenta by studying the action (2.1). Usually, they are defined as time-derivatives of fields. This definition also remains in the case of vector fields, with one momenta existing for each space-time component of A_μ. In particular, they are given by:

$$\Pi^0 = 0 \qquad \text{and} \qquad \Pi^i = F_{0i}\,. \tag{2.4}$$

We can notice that the temporal component of the conjugated momenta is equal to zero. This is known as the primary constraint. Let us now define Poisson brackets of two functions f and g as

$$\{f, g\} = \int d^3x \left(\frac{\delta f}{\delta A_\mu(x)} \frac{\delta g}{\delta \Pi^\mu(x)} - \frac{\delta g}{\delta A_\mu(x)} \frac{\delta f}{\delta \Pi^\mu(x)} \right). \tag{2.5}$$

Then, it follows that:

$$\left\{ A_\mu(\boldsymbol{x}, t), \Pi^\nu(\boldsymbol{y}, t) \right\} = \delta^\nu_\mu \delta\left(\boldsymbol{x} - \boldsymbol{y}\right). \tag{2.6}$$

Therefore, we can see that in this formalism, the Poisson brackets of Π^0 and A^0 are non-vanishing despite the relation (2.4). However, following [103], this constraint should simply then be carefully treated, and set to zero only after one computes the Poisson brackets. Otherwise, we will find a wrong result. This is known as the "weak equation":

$$\Pi^0 \approx 0. \tag{2.7}$$

The total Hamiltonian is given by:

$$H = \frac{1}{2}\int d^3x \left(\Pi^i\Pi^i + 2\Pi^i A_{0,i} - m^2 A_0 A_0 - A_{i,i}A_{j,j} + A_{i,j}A_{i,j} \right.$$
$$\left. + m^2 A_i A_i + v\Pi_0\right), \tag{2.8}$$

where v is the Lagrange multiplier. Then, for a function f, the equations of motion are given by

$$\dot{f} = \{f, H\}. \tag{2.9}$$

Clearly, the primary constraint has to satisfy

$$\dot{\Pi}^0 = 0. \tag{2.10}$$

Then, using (2.9), from the above condition we obtain another constraint

$$\Phi = \Pi^i_{,i} + m^2 A_0 = 0 \tag{2.11}$$

refereed to as the secondary constraint. Using the consistency condition

$$\dot{\Phi} = 0, \tag{2.12}$$

we can easily see that (2.7) and (2.11) are the only constraints in this theory—(2.12) determines the Lagrange multiplier v. By definition, since the Poisson brackets of these two constraints are non-vanishing:

$$\left\{\Phi(\boldsymbol{x}, t), \Pi^0(\boldsymbol{y}, t)\right\} = m^2\delta\left(\boldsymbol{x} - \boldsymbol{y}\right), \tag{2.13}$$

they are known as the second class. This means that we can set them to zero and express the Hamiltonian just in terms of the canonical variables, A_i and Π^i, dropping the Π^0 and A_0 completely from consideration [103].[1] The solution of (2.11) is given by

$$A_0 = -\frac{1}{m^2}\Pi^i_{,i}. \tag{2.14}$$

[1] Of course, one has to reintroduce A_0 if interested in the manifestly covariant formulation of the theory. However, in this case, one can simply define it through (2.14).

By substituting it in (2.8), we find:

$$H = \frac{1}{2}\int d^3x \left(\Pi^i\Pi^i + \frac{1}{m^2}\Pi^i_{,i}\Pi^j_{,j} - A_{i,i}A_{j,j} + A_{i,j}A_{i,j} + m^2 A_i A_i\right). \tag{2.15}$$

The Hamiltonian equations of motion are then given by

$$\dot{\Pi}^i = \left\{\Pi^i, H\right\} \qquad \text{and} \qquad \dot{A}_i = \{A_i, H\}. \tag{2.16}$$

The number of degrees of freedom per point in space is defined as number of initial conditions divided by two [32]. Therefore, we can find the first crucial difference between massive and massless photon—Proca theory has three degrees of freedom, while Maxwell theory has only two of them. Let us analyse now these degrees of freedom in detail.

2.1.1.2 The Physical Degrees of Freedom

Although the above procedure was insightful on the type of constraints that appear in the theory and has allowed us to express the Hamiltonian just in terms of the canonical variables, it can become rather lengthy, especially once one considers self-interacting fields. Therefore, here we will present an alternative but equivalent method that expresses the Lagrangian in terms of the physical degrees of freedom—the transverse and longitudinal modes. Such an approach was also considered in [33, 45, 104].

First, we will return to the initial action (2.1), separate the temporal and spatial parts of the vector field, and decompose the spatial part as

$$A_i = A_i^T + \chi_{,i}, \qquad \text{where} \qquad A^T_{i,i} = 0. \tag{2.17}$$

We will refer to the A_i^T component as the transverse modes, while the field χ is the longitudinal mode. The action (2.1) then becomes

$$\begin{aligned} S = \frac{1}{2}\int d^4x \Big[&A_0\left(-\Delta + m^2\right)A_0 + 2A_0\Delta\dot{\chi} - \left(\dot{\chi}\Delta\dot{\chi} - m^2\chi\Delta\chi\right) \\ &+ \left(\dot{A}_i^T\dot{A}_i^T - A^T_{i,j}A^T_{i,j} - m^2 A_i^T A_i^T\right)\Big]. \end{aligned} \tag{2.18}$$

As there are no time derivatives acting on the A_0 component, we can rewrite the theory in such a way that it contains only the longitudinal and transverse modes. Varying the action with respect to A_0, we find that it satisfies the following constraint:

$$\left(-\Delta + m^2\right)A_0 = -\Delta\dot{\chi}, \tag{2.19}$$

whose solution is given by

$$A_0 = \frac{-\Delta}{-\Delta + m^2}\dot{\chi}. \tag{2.20}$$

The operator acting on the longitudinal modes is to be understood in terms of Fourier modes. If we were to express the longitudinal modes in the following way

$$\chi = \int d^3k \tilde{\chi}(\boldsymbol{k}) e^{i\boldsymbol{k}\boldsymbol{x}}, \tag{2.21}$$

the operator becomes

$$\frac{-\Delta}{-\Delta + m^2}\dot{\chi} = \int d^3k \frac{|\boldsymbol{k}|^2}{|\boldsymbol{k}|^2 + m^2}\dot{\tilde{\chi}}(\boldsymbol{k}) e^{i\boldsymbol{k}\boldsymbol{x}}. \tag{2.22}$$

Substituting (2.20) back into the action, we obtain

$$S = -\frac{1}{2}\int d^4x \left[A_i^T\left(\Box + m^2\right)A_i^T + \chi\left(\Box + m^2\right)\frac{m^2(-\Delta)}{-\Delta + m^2}\chi\right]. \tag{2.23}$$

Thus, the action is now expressed only in terms of the propagating degrees of freedom—the longitudinal and transverse modes. However, we can notice that the longitudinal modes are not canonically normalised. Defining the normalised ones as

$$\chi_n = m\sqrt{\frac{-\Delta}{-\Delta + m^2}}\chi, \tag{2.24}$$

we obtain

$$S = \frac{1}{2}\int d^4x \left[\dot{A}_i^T\dot{A}_i^T - A_{i,j}^T A_{i,j}^T - m^2 A_i^T A_i^T + \dot{\chi}_n\dot{\chi}_n - \chi_{n,i}\chi_{n,i} - m^2\chi_n\chi_n\right]. \tag{2.25}$$

We can see that in comparison to the manifestly covariant form of the theory, given by (2.1), this manifestly non-covariant form greatly simplifies the form of the equations of motion.

One the one hand, in the manifestly covariant formulation, the vector field A^μ satisfies the following equations

$$F^{\mu\nu}_{\ ,\mu} + m^2 A^\nu = 0. \tag{2.26}$$

From here, it follows that

$$A^\nu_{\ ,\nu} = 0, \tag{2.27}$$

which substituting back into (2.26) reduces it to

$$\left(\Box + m^2\right) A^\nu = 0. \tag{2.28}$$

If not careful enough, this form of the equation of motion could be misleading—one could conclude that even the A_0 component of the vector field is propagating. However, we know that this is not the case—to count the degrees of freedom, one should then also take into account (2.27).

In the manifestly non-covariant formulation, on the other hand, the only equations that we have to concern ourselves with are the equations of motion for the transverse and longitudinal modes:

$$\left(\Box + m^2\right) A_i^T = 0 \qquad \text{and} \qquad \left(\Box + m^2\right) \chi_n = 0, \tag{2.29}$$

whose solutions are plane waves. Therefore, this formulation allows us to examine the degrees of freedom directly.

2.1.1.3 Canonical Quantization of the Theory

The formulation of the theory in terms of propagating modes makes its quantization straight-forward. As a first step, we find the conjugated momenta associated with the longitudinal and transverse modes:

$$\pi^{Ti} = \dot{A}_i^T \qquad \text{and} \qquad \pi_n = \dot{\chi}_n. \tag{2.30}$$

Then, we can easily obtain the Hamiltonian density:

$$\begin{aligned} \mathcal{H} &= \mathcal{H}_{A^T} + \mathcal{H}_{\chi_n}, \qquad \text{where} \\ \mathcal{H}_{A^T} &= \frac{1}{2}\left(\pi_i^T \pi_i^T + A_{i,j}^T A_{i,j}^T + m^2 A_i^T A_i^T\right) \qquad \text{and} \\ \mathcal{H}_{\chi_n} &= \frac{1}{2}\left(\pi_n^2 + \chi_{n,i}\chi_{n,i} + m^2 \chi_n^2\right). \end{aligned} \tag{2.31}$$

We may wonder how does this expression connect with (2.15). For that we just need the decomposition of the canonical momenta. It is given by

$$\Pi^i = \pi^{Ti} + \frac{1}{-\Delta}\pi_{\chi,i} \tag{2.32}$$

where $\pi_{\chi,i}$ is the canonical momenta associated with the original longitudinal mode χ, and is connected with the normalised longitudinal momenta by

$$\pi_n = \sqrt{\frac{-\Delta + m^2}{-\Delta m^2}}\pi_\chi. \tag{2.33}$$

Therefore, we can see that the quantisation of this theory is straight-forward generalisation of that of a scalar field. In order to do it, we promote the transverse and longitudinal modes, and their respective conjugated momenta into operators and postulate equal-time canonical commutation relations:

$$\begin{aligned}&\left[\hat{\chi}_n(\boldsymbol{x},t),\hat{\pi}_n(\boldsymbol{y},t)\right]=i\delta\left(\boldsymbol{x}-\boldsymbol{y}\right)\\&\left[\hat{\chi}_n(\boldsymbol{x},t),\hat{\chi}_n(\boldsymbol{y},t)\right]=0\qquad\qquad\left[\hat{\pi}_n(\boldsymbol{x},t),\hat{\pi}_n(\boldsymbol{y},t)\right]=0\end{aligned}\tag{2.34}$$

and

$$\begin{aligned}&\left[\hat{A}_i^T(\boldsymbol{x},t),\hat{\pi}^{Tj}(\boldsymbol{y},t)\right]=i\left(\delta_i^j+\frac{\partial_i\partial^j}{\Delta}\right)\delta\left(\boldsymbol{x}-\boldsymbol{y}\right)\\&\left[\hat{A}_i^T(\boldsymbol{x},t),\hat{A}_j^T(\boldsymbol{y},t)\right]=0\qquad\qquad\left[\hat{\pi}^{Ti}(\boldsymbol{x},t),\hat{\pi}^{Tj}(\boldsymbol{y},t)\right]=0.\end{aligned}\tag{2.35}$$

In the first relation, both partial derivatives can be performed either with respect to $\boldsymbol{x}$ or with respect to $\boldsymbol{y}$. We can expand the operators in terms of the time-independent creation and annihilation operators:

$$\begin{aligned}&\hat{\chi}_n(\boldsymbol{x},t)=\int\frac{d^3k}{(2\pi)^{\frac{3}{2}}}\frac{1}{\sqrt{2\omega_{\mathbf{k}}}}\left(e^{-ikx}\hat{\hat{a}}_{\mathbf{k}}^{-}+e^{ikx}\hat{\hat{a}}_{\mathbf{k}}^{+}\right),\qquad\text{and}\\&\hat{A}^{Ti}(\boldsymbol{x},t)=\sum_{\sigma=1,2}\int\frac{d^3k}{(2\pi)^{\frac{3}{2}}}\frac{1}{\sqrt{2\omega_{\mathbf{k}}}}\epsilon_{\mathbf{k}\sigma}^i\left(e^{-ikx}\hat{a}_{\mathbf{k}\sigma}^{-}+e^{ikx}\hat{a}_{\mathbf{k}\sigma}^{+}\right).\end{aligned}\tag{2.36}$$

where, in the exponent, the product of the 4-vectors is given by $kx=k_\mu x^\mu$, $\omega_k^2=\boldsymbol{k}^2+m^2$ is the frequency and $\epsilon_{\mathbf{k}\sigma}^i$ are the two transverse polarisation vectors, that satisfy:

$$\epsilon_{\mathbf{k}\sigma}\epsilon_{\mathbf{k}\sigma'}=\delta_{\sigma\sigma'},\qquad \boldsymbol{k}\cdot\epsilon_{\mathbf{k}\sigma}=0\qquad\text{and}\qquad\sum_{\sigma=1}^{2}\epsilon_{\mathbf{k}\sigma}^i\epsilon_{\mathbf{k}\sigma}^j=\delta^{ij}-\frac{k^ik^j}{|\boldsymbol{k}|^2}.\tag{2.37}$$

Using the commutation relations (2.34) and (2.35), one can show that the annihilation and creation operators for the longitudinal modes satisfy following relations:

$$\left[\hat{\hat{a}}_{\mathbf{k}}^{-},\hat{\hat{a}}_{\mathbf{k}'}^{+}\right]=\delta\left(\boldsymbol{k}-\boldsymbol{k}'\right)\qquad\left[\hat{\hat{a}}_{\mathbf{k}}^{-},\hat{\hat{a}}_{\mathbf{k}'}^{-}\right]=0\qquad\left[\hat{\hat{a}}_{\mathbf{k}}^{+},\hat{\hat{a}}_{\mathbf{k}'}^{+}\right]=0\tag{2.38}$$

while the ones for the transverse modes satisfy:

$$\left[\hat{a}_{\mathbf{k}\sigma}^{-},\hat{a}_{\mathbf{k}\sigma'}^{+}\right]=\delta_{\sigma\sigma'}\delta\left(\boldsymbol{k}-\boldsymbol{k}'\right)\qquad\left[\hat{a}_{\mathbf{k}\sigma}^{-},\hat{a}_{\mathbf{k}\sigma'}^{-}\right]=0\qquad\left[\hat{a}_{\mathbf{k}\sigma}^{+},\hat{a}_{\mathbf{k}\sigma'}^{+}\right]=0\tag{2.39}$$

Inserting them into the Hamiltonian operator we obtain for the longitudinal modes:

$$\hat{H}_{\chi_n} = \int d^3k \left(\omega_{\mathbf{k}} \hat{a}^+_{\mathbf{k}} \hat{a}^-_{\mathbf{k}} + \frac{\omega_{\mathbf{k}}}{2} \delta^{(0)} \right) \tag{2.40}$$

We can notice that the vacuum energy

$$E_{\chi_n} = \langle 0 | \hat{H}_{\chi_n} | 0 \rangle = \int d^3k \frac{1}{2} \omega_{\mathbf{k}} \delta^{(0)}, \tag{2.41}$$

where $|0\rangle$ is the vacuum—state with minimal energy that is annihilated by the annihilation operator $\hat{a}\,|0\rangle = 0$—is clearly divergent. This is first, due to the integration over the whole space-time that reflects in the presence of a delta function, and second because the field is a collection of infinitely many harmonic oscillators. Nevertheless, as we are in a flat space we can neglect both divergences—only the difference between two energies can be measured by an experiment—and hence redefine the Hamiltonian as

$$\hat{H}_{\chi_n} = \int d^3k \omega_{\mathbf{k}} \hat{a}^+_{\mathbf{k}} \hat{a}^-_{\mathbf{k}} \tag{2.42}$$

for the longitudinal modes, while for the transverse ones we find:

$$\hat{H}_{A^T} = \int d^3k \sum_{\sigma=1,2} \omega_{\mathbf{k}} \hat{a}^+_{\mathbf{k}\sigma} \hat{a}^-_{\mathbf{k}\sigma} \tag{2.43}$$

2.1.1.4 The Propagators

One of the key elements in the upcoming analysis that will be the propagators associated with the fields. Let us first calculate them for the longitudinal and transverse modes. In the position space, the propagator for the longitudinal modes is defined by the following relation:

$$\begin{aligned} \Delta_{\chi_n}(x-y) &= \langle 0 | T \chi_n(x) \chi_n(y) | 0 \rangle \\ &= \theta\left(x^0 - y^0\right) D(x-y)_{\chi_n} + \theta\left(y^0 - x^0\right) D(y-x)_{\chi_n}, \end{aligned} \tag{2.44}$$

where

$$D_{\chi_n}(x-y) \equiv \langle 0 | \chi_n(x) \chi_n(y) | 0 \rangle = \int \frac{d^3k}{(2\pi)^3} \frac{1}{2\omega_{\mathbf{k}}} e^{-ik(x-y)}. \tag{2.45}$$

Here we have used (2.36) to obtain the last equality. Then, using the definition (2.44), we obtain

$$\Delta_{\chi_n}(x-y) = \int \frac{d^4k}{(2\pi)^4} \frac{i}{k^2 - m^2 + i\epsilon} e^{-ik(x-y)}. \tag{2.46}$$

In the case of the transverse modes, we must additionally use the completeness relation (2.37). Then, we find that the propagator for the transverse modes is given by:

$$\Delta^T_{ij}(x-y) = \int \frac{d^4k}{(2\pi)^4} \frac{i}{k^2 - m^2 + i\epsilon} \left(\delta_{ij} - \frac{k_i k_j}{|\boldsymbol{k}|^2}\right) e^{-ik(x-y)}. \tag{2.47}$$

So far, we have studied the physical degrees of freedom. The form of this theory is manifestly non-covariant. Thus, let us see how to connect these results with the manifestly covariant formulation.

2.1.1.5 Returning to the Manifestly Covariant Form

In order to collect the results that we have previously found into the manifestly covariant form, we will first consider the polarisation vectors

$$\varepsilon^\mu_\sigma, \qquad \sigma = 1, 2, 3 \tag{2.48}$$

We can choose the transverse polarisation vectors to be of the following form

$$\varepsilon^\mu_1 = (0, \boldsymbol{\epsilon}_{\mathbf{k}1}) \qquad \text{and} \qquad \varepsilon^\mu_2 = (0, \boldsymbol{\epsilon}_{\mathbf{k}2}), \tag{2.49}$$

where $\boldsymbol{\epsilon}_{\mathbf{k}1}$ and $\boldsymbol{\epsilon}_{\mathbf{k}2}$ are the 3-transverse polarisation vectors given in (2.37). The third polarisation vector that accounts for the longitudinal modes can them be chosen as [102]

$$\varepsilon^\mu_3 = \left(\frac{|\boldsymbol{k}|}{m}, \frac{\boldsymbol{k}k_0}{|\boldsymbol{k}|m}\right). \tag{2.50}$$

All three polarisation vectors are normalised

$$\varepsilon^\mu_\sigma \varepsilon_{\mu\lambda} = -\delta_{\sigma\lambda}, \tag{2.51}$$

have the following property

$$k_\mu \varepsilon^\mu_\sigma = 0, \tag{2.52}$$

and satisfy the completeness relation:

$$\sum_{\sigma=1}^{3} \varepsilon_{\mu\sigma}\varepsilon_{\nu\sigma} = -\eta_{\mu\nu} + \frac{k_\mu k_\nu}{m^2}. \tag{2.53}$$

Then, we can collect the oscillator expansion of the modes to the following expression:

$$\hat{A}^{\mu}(x) = \sum_{\sigma=1}^{3} \int \frac{d^3k}{(2\pi)^{\frac{3}{2}}} \frac{1}{\sqrt{2\omega_{\mathbf{k}}}} \varepsilon^{\mu}_{\mathbf{k}\sigma} \left(e^{-ikx} \hat{a}^{-}_{\mathbf{k}\sigma} + e^{ikx} \hat{a}^{+}_{\mathbf{k}\sigma} \right). \tag{2.54}$$

Here, $\hat{A}^0$ operator is defined through (2.20), and creation and annihilation operators of the normalised longitudinal modes are related with the ones previously given as

$$\hat{a}^{-}_{\mathbf{k}3} = -i\hat{\tilde{a}}^{-}_{\mathbf{k}} \qquad \text{and} \qquad \hat{a}^{+}_{\mathbf{k}3} = i\hat{\tilde{a}}^{+}_{\mathbf{k}}. \tag{2.55}$$

Furthermore, the propagator of the theory in the momentum space can be written in the manifestly covariant form as

$$\tilde{\Delta}_{\mu\nu}(k) = \left(-\eta_{\mu\nu} + \frac{k_\mu k_\nu}{m^2} \right) \frac{i}{k^2 - m^2 + i\epsilon}. \tag{2.56}$$

Therefore, we can notice something curious—the above propagator has a term that is singular in mass. Let us now see how these results are generalised for the case of a non-Abelian vector field.

2.1.2 *Massive Yang-Mills Theory*

In this work, we will be interested in the simplest case of massive non-Abelian vector fields—the SU(2) theory. Its action is given by

$$S = \int d^4x \mathrm{Tr} \left[-\frac{1}{2} F_{\mu\nu} F^{\mu\nu} + m^2 A_\mu A^\mu \right]. \tag{2.57}$$

In contrast to the Abelian case, the field strength tensor is now given by

$$F_{\mu\nu} = D_\mu A_\nu - D_\nu A_\mu, \tag{2.58}$$

where D_μ is the covariant derivative,

$$D_\mu = \partial_\mu + igA_\mu, \tag{2.59}$$

and g is the coupling constant. Throughout this work, we will assume that

$$g \ll 1. \tag{2.60}$$

The vector fields A_μ are 2×2 hermitian matrices. They can be expanded in terms of the generators of SU(2) as

$$A_\mu = A^a_\mu T^a \quad , \quad T^a = \frac{\sigma^a}{2} \quad , \quad a = 1, 2, 3, \tag{2.61}$$

where σ^a are the Pauli matrices. The generators satisfy the following relations:

$$\mathrm{Tr}\left(T^a T^b\right) = \frac{\delta^{ab}}{2}, \qquad \text{and} \qquad \left[T^a, T^b\right] = i\varepsilon^{abc} T^c, \tag{2.62}$$

where ε^{abc} is the Levi-Civita symbol. Expanding the vector field in terms of the generators, we can show that the field strength becomes

$$F^a_{\mu\nu} = A^a_{\nu,\mu} - A^a_{\mu,\nu} - g\varepsilon^{abc} A^b_\mu A^c_\nu. \tag{2.63}$$

Therefore, in comparison to the Proca theory, massive Yang-Mills theory has cubic and quartic self-interactions:

$$\begin{aligned} S = \int d^4x \Bigg(& -\frac{1}{4} F^a_{\mu\nu} F^{a\mu\nu} + \frac{m^2}{2} A^a_\mu A^{a\mu} \\ & + g\varepsilon^{abc} A^{a\nu,\mu} A^b_\mu A^c_\nu - \frac{g^2}{4} \varepsilon^{abc} \varepsilon^{ade} A^b_\mu A^c_\nu A^{d\mu} A^{e\nu} \Bigg). \end{aligned} \tag{2.64}$$

In addition, we can see that by setting the coupling constant to zero we will obtain just three Proca theories, one for each $a = 1, 2, 3$. Therefore, in the free case, we can easily generalize the results that we have previously found for the Proca field. All we have to do is include an additional color index. In particular, the propagator in the momentum space is given by

$$\tilde{\Delta}^{ab}_{\mu\nu}(k) = \left(-\eta_{\mu\nu} + \frac{k_\mu k_\nu}{m^2}\right) \frac{i\delta^{ab}}{k^2 - m^2 + i\epsilon}. \tag{2.65}$$

We can notice that the propagator in both Proca and massive Yang-Mills theory is singular in mass. This suggests that the theories might have a discontinuity in the massless limit. Let us investigate what is the source of this behavior. First, we can find the propagator for the original longitudinal modes. Using (2.33) and (2.46), we can find that in the momentum space, it is given by

$$\tilde{\Delta}_\chi(k) = \frac{\mathbf{k}^2 + m^2}{m^2 \mathbf{k}^2} \frac{i}{k^2 - m^2 + i\epsilon}. \tag{2.66}$$

Now, let us consider the purely spatial part of (2.56). It is easy to show that it can be rewritten as

$$
\begin{aligned}
\tilde{\Delta}_{ij}(k) &= \left(\delta_{ij} - \frac{k_i k_j}{|\boldsymbol{k}|^2} + k_i k_j \frac{|\boldsymbol{k}|^2 + m^2}{m^2 |\boldsymbol{k}|^2}\right) \frac{i}{k^2 - m^2 + i\epsilon} \\
&= \tilde{\Delta}^T(k) + k_i k_j \tilde{\Delta}_\chi(k),
\end{aligned}
\tag{2.67}
$$

where $\tilde{\Delta}^T(k)$ is the propagator of the transverse modes in momentum space. Therefore, we can see that the singular-in-mass behaviour of the propagator (2.56) is purely due to the longitudinal modes—degrees of freedom that are not present in the massless theory.

2.1.3 Interacting Proca Theory

In Proca theory, the propagator is not problematic if the theory is coupled through a conserved source. The exchange of photons between two sources is given by

$$
\int d^4x \int d^4y J^\mu(x) \Delta_{\mu\nu}(x - y) J^\nu(y). \tag{2.68}
$$

Therefore, the divergent term in the propagator drops out because the current is conserved—due to it, the longitudinal mode decouples from the remaining degrees of freedom.

In the massive Yang-Mills theory, the situation, however, is much more complicated—in addition to the source, this theory also has self-interactions. As a result, we will find that these divergences remain.

In this chapter, our goal is to understand these divergences and follow the methods that have led to the conclusion that the massless limit of the massive Yang-Mills theory is not smooth. At first, we will simplify this analysis, going through the basics with the toy model of interacting Proca theory.

As we have seen, the divergences are not problematic in the presence of the conserved source. However, following the line of the massive Yang-Mills theory, we can expect that the same will not be the case if we consider the self-interacting theory. Therefore, the simplest modification to Proca theory that will give a discontinuity in the massless limit is given by

$$
S = \int d^4x \left(-\frac{1}{4} F_{\mu\nu} F^{\mu\nu} + \frac{m^2}{2} A_\mu A^\mu - \frac{g^2}{2} A_\mu A_\nu A^{\nu,\mu}\right). \tag{2.69}
$$

This theory arises in the vector Galileon theories [105]. Here, we will use it as a toy model that will help us analyze the key properties of massive vector theories and allow us to construct the insightful methods that we will later use in the massive Yang-Mills theory.

2.2 Unitarity and the Cutting Rules

Let us consider a physical process, such as a scattering of two particles. Then we can define the *in-state* $|a\rangle_{in}$—state of the system at time $t = -\infty$—that we can characterise with the moments and spins of these particles. Naturally, we are then curious about the state of the system at a later time. An *out-state* $|a\rangle_{out}$ is the state of the system at time $t = +\infty$. Both in and out states correspond to non-interacting particles. However, if in the intermediate time these particles interact, the resulting state can differ from the initial one, and we can find the transition probability for such a process to occur.

S-matrix—a unitary operator that transforms an in-state into an out-state—provides a connection between the formal theory and observations. Following the standard conventions [106], it is defined as

$$|a\rangle_{out} = \hat{S}^{\dagger}\,|a\rangle_{in}\,. \tag{2.70}$$

If there is no scattering, this operator is just an identity operator. Therefore, to single out the physical process, we can define a T-matrix by

$$\hat{S} = \hat{I} + i\hat{T}, \tag{2.71}$$

where $\hat{I}$ is the identity operator. The T matrix satisfies the following identity

$$i\left(\hat{T} - \hat{T}^{\dagger}\right) = -\hat{T}^{\dagger}\hat{T}, \tag{2.72}$$

that follows from the unitarity of the S-matrix, given by

$$\hat{S}^{\dagger}\hat{S} = \hat{S}\hat{S}^{\dagger} = \hat{I}. \tag{2.73}$$

Yet, the unitarity is given by the formal construction of the S-matrix. As pointed out in [106], when considering an interacting theory, nothing guarantees that the relation (2.73) truly holds. Rather, we should confirm that it is satisfied by each particular theory that we are considering.

The Cutkosky rules developed in [107] provide us with this opportunity [108]. They relate the Feynman diagrams with the unitarity of the S-matrix. In this chapter, we will outline this method, by closely following the approach of [106]. Applying it to the toy model introduced in (2.69), and studying the corrections to the propagator, we will find that this method suggests that (2.69) violates unitarity.

2.2.1 The Feynman Rules for the Toy Model

The simplest example on which we can analyse the unitarity of the model (2.69) are the corrections to the propagator of the vector modes. For that, we first need to find the Feynman rules. It can be easily shown that in the momentum space, they are given by:

$$\Delta_{\mu\nu}(k) = \left(-\eta_{\mu\nu} + \frac{k_\mu k_\nu}{m^2}\right)\frac{i}{k^2 - m^2}$$

$$ig^2 2V^{\mu\nu\gamma}(k, p, q)$$

$$V^{\mu\nu\gamma}(k, p, q) = i\left(k^\mu \eta^{\nu\gamma} + p^\nu \eta^{\mu\gamma} + q^\gamma \eta^{\mu\nu}\right),$$

with all moments outgoing in the vertex, and correspond to the following diagrams (Figs. 2.1 and 2.2):

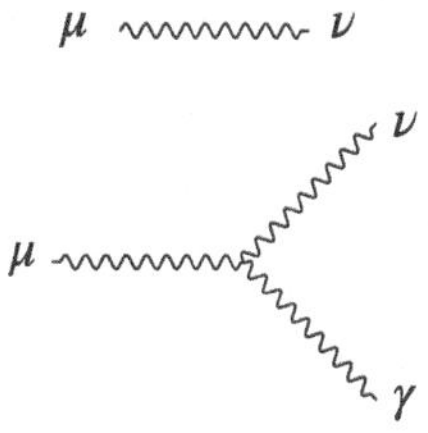

Fig. 2.1 The propagator and the vertex for the Feynman rules

Then, the diagrams contributing to the propagator of the massive photon up to two-loop corrections are given by:

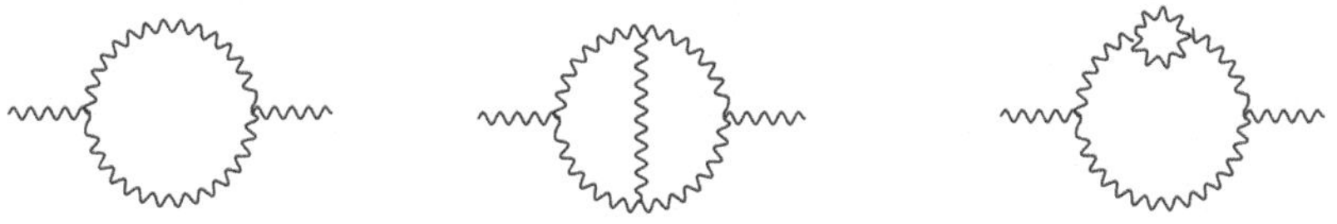

Fig. 2.2 Two-loop corrections to the propagator in the toy model

2.2.2 *The Circles and Cuts*

Let us now find what can we learn about the unitarity of the toy model from the diagrams given above. First, we can notice that the propagator (2.56) can be written in the position space as

$$\Delta_{\mu\nu}(x - y) = \theta\left(x^0 - y^0\right)\Delta^+_{\mu\nu}(x - y) + \theta\left(y^0 - x^0\right)\Delta^-_{\mu\nu}(x - y), \qquad (2.74)$$

where θ is the Heaviside step function, and

$$\Delta^\pm_{\mu\nu}(x - y) = \int \frac{d^4k}{(2\pi)^3}\left(-\eta_{\mu\nu} + \frac{k_\mu k_\nu}{m^2}\right)\theta\left(\pm k^0\right)\delta\left(k^2 - m^2\right). \qquad (2.75)$$

Largest time equations—expressions that generalise (2.74) to any diagram—then follow from a few simple rules [106]. In addition to the Feynman rules above, following [106], let us add an additional set of rules that concern the existence of a circle at a vertex:

1. If the vertex has a circle on it, we will multiply the expression by (-1).
2. If the line has no circles, we keep $\Delta_{\mu\nu}(x-y)$.
3. If the line starts with a circle and ends with none, we replace $\Delta_{\mu\nu}(x-y)$ with $\Delta^{-}_{\mu\nu}(x-y)$.
4. If a line ends with a circle and begins with none, we replace $\Delta_{\mu\nu}(x-y)$ with $\Delta^{+}_{\mu\nu}(x-y)$.
5. If both ends of a line are circled, we replace $\Delta_{\mu\nu}(x-y)$ with its complex conjugate $\Delta^{*}_{\mu\nu}(x-y)$.

Then, the largest time equation for the 1-loop correction to the propagator is depicted by the following diagrams:

Fig. 2.3 The largest time equation

We can easily see that this relation holds—both first and the last diagrams cancel with each other, as well as the middle ones if $x_2^0 > x_1^0$. Otherwise, the first will cancel the third one, while the second cancels the last one. In particular, we can notice that the sum over all possible circles (excluding the external lines) vanishes—this will be the case for each diagram.

Let us now consider the momentum space, apply the conservation of energy and momentum at each vertex and assume that the incoming external lines indicate incoming positive energy flow. Due to the dependence of the theta-function on energy in (2.75), positive energy flows only towards a circle. If we now have several circles in one region, like in the following diagram:

Fig. 2.4 An example of a diagram with circles

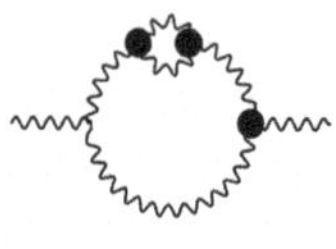

then we can replace these regions with a cut in the following way:

Fig. 2.5 Replacement of circles with a cut

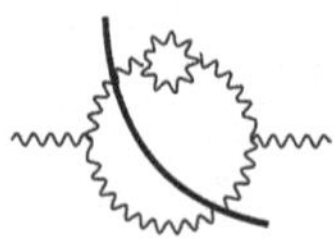

The positive energy flow will always be towards the right. In other words, the sum over all possible circles becomes the sum over all possible cuts. In particular, the Fig. 2.3, becomes

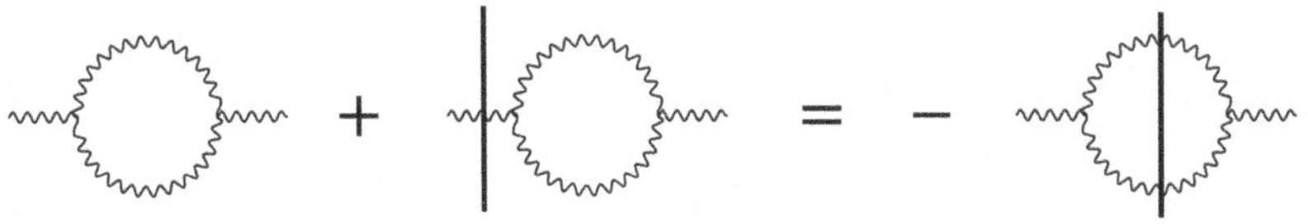

Fig. 2.6 Replacement of the 1-loop corrections with cuts

We can notice that we have left out one diagram—the third diagram of Fig. 2.3 is zero if the first of the external lines represents an incoming line. Otherwise, we would need an additional circle at the second vertex for the energy to flow (Figs. 2.4, 2.5, and 2.6).

We can write this result in a formal way, as presented in [106]. If $F(x_1, \ldots, x_n)$ represents a diagram with n vertices, we have the following identity, known as the cutting equation:

$$F(x_1, \ldots, x_n) + F^*(x_1, \ldots, x_n) = -\sum_{cuts} F(x_1, \ldots, x_n) \tag{2.76}$$

Then, if $|a\rangle$, $|b\rangle$, $|c\rangle$ denote the states of the system, we can see that the unitarity condition (2.72), rewritten in the following way

$$i \langle b| \hat{T} - \hat{T}^\dagger |a\rangle = -\sum_c \langle b| \hat{T}^\dagger |c\rangle \langle c| \hat{T} |a\rangle, \tag{2.77}$$

where we have inserted a complete set of states for $|c\rangle$, is just the cutting equation—the terms on the left hand side are just the imaginary part of the diagram, while the right hand side corresponds to the cuts of this diagram.

Thus, the procedure above introduces another set of rules—the Cutkotsky rules—that are momentum space analogous to the ones involving circles:

1. If a line does not have a cut through it, we write the standard propagator (2.56).
2. If a line cutted through with positive energy flow, one replaces the propagator in the momentum space (2.56) with

$$\Delta^+_{\mu\nu}(k) = 2\pi \left(-\eta_{\mu\nu} + \frac{k_\mu k_\nu}{m^2}\right) \theta\left(k^0\right) \delta\left(k^2 - m^2\right) \tag{2.78}$$

3. For a cut with the negative energy flow, one replaces the propagator with

$$\Delta^-_{\mu\nu}(k) = 2\pi \left(-\eta_{\mu\nu} + \frac{k_\mu k_\nu}{m^2}\right) \theta\left(-k^0\right) \delta\left(k^2 - m^2\right) \tag{2.79}$$

Let us now apply them to calculate the propagator corrections. From now on, all of the external momenta will be denoted by p (Figs. 2.7 and 2.8).

2.2.2.1 The One-Loop Corrections

In this case only one diagram provides a contribution

Fig. 2.7 The one-loop corrections to the propagator in the toy model

Since the incoming and outgoing particles are on-shell, we have $p_\mu \varepsilon_\sigma^\mu (p) = 0$, where ε_σ^μ are the polarisation vectors that satisfy the completeness relation (2.53). The imaginary part is given by the sum of all possible cuts:

$$\mathrm{Im}\,(\Gamma_1) = \frac{1}{2} \sum_{cuts} (i\Gamma_1) \tag{2.80}$$

The above diagram has only one cut

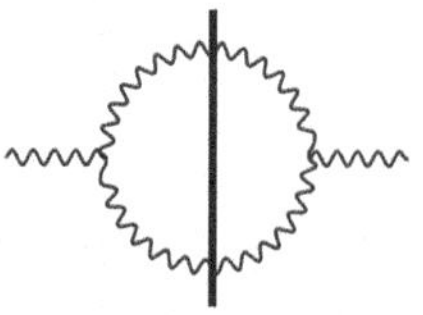

Fig. 2.8 The one-loop corrections to the propagator in the toy model

Using the following trick [41]

$$\int \frac{d^4k}{(2\pi)^4} = \int \frac{d^4k}{(2\pi)^4} \frac{d^4q}{(2\pi)^4} (2\pi)^4 \delta(q - p + k), \tag{2.81}$$

we can show that it's imaginary part is given by

$$\begin{aligned} \mathrm{Im}\,(\Gamma_1) = \varepsilon_\mu (p) \varepsilon_\nu (-p) \int \frac{d^4k}{(2\pi)^3} \int \frac{d^4q}{(2\pi)^3} (2\pi)^4 \delta(q + k - p) \times \\ \times \theta(k_0) \theta(q_0) \delta\left(k^2 - m^2\right) \delta\left(q^2 - m^2\right) L^{\mu\nu}, \end{aligned} \tag{2.82}$$

where

$$L^{\mu\nu} = \frac{g^4}{16} V^{\mu\alpha\beta}(-p, k, q) \left(-\eta_{\alpha\gamma} + \frac{k_\alpha k_\gamma}{m^2} \right) \times \\ \times \left(-\eta_{\beta\delta} + \frac{q_\beta q_\delta}{m^2} \right) V^{\gamma\delta\nu}(-k, -q, p) \,. \tag{2.83}$$

Because of the delta functions, the internal momenta are on-shell, i.e. $k^2 = q^2 = m^2$. Then, noting that

$$k^\gamma \left(-\eta_{\alpha\gamma} + \frac{k_\alpha k_\gamma}{m^2} \right) = 0 \qquad \text{for} \qquad k^2 = m^2, \tag{2.84}$$

it follows that $L_{\mu\nu} = 0$. Thus, at one loop there is no discontinuity. Nonetheless, the same situation does not hold for two-loop corrections (Figs. 2.9 and 2.10).

2.2.2.2 The Two-Loop Corrections

For simplicity, here we will consider only 3-particle cuts of the 2-loop diagrams. In this case, the following diagrams have a non-vanishing imaginary part

Fig. 2.9 The two-loop corrections to the propagator in the toy model

The 3-particle cuts of these diagrams are given by

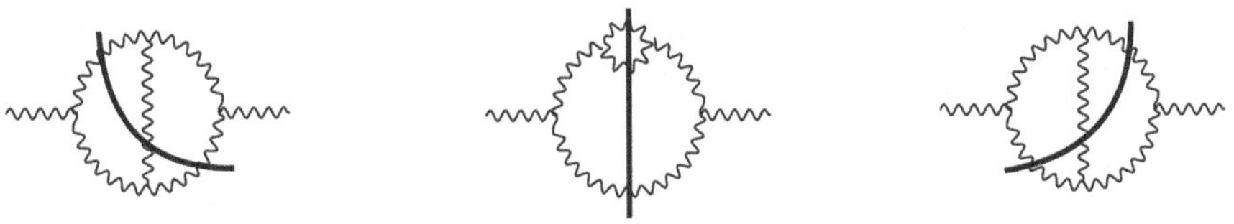

Fig. 2.10 The cuts of the two-loop corrections to the propagator in the toy model

Let us first consider the middle diagram. Using the generalisation of the trick for the 1-loop contributions

$$\int \frac{d^4k}{(2\pi)^4} \int \frac{d^4l}{(2\pi)^4} = \int \frac{d^4k}{(2\pi)^4} \int \frac{d^4l}{(2\pi)^4} \int d^4q \delta\,(p-k-l-q)\,, \tag{2.85}$$

and the Cutkotsky rules, we find that the 3-particle cut contribution to the imaginary part is given by

$$\begin{aligned}\mathrm{Im}\,(\Gamma_{2A}) = &\int \frac{d^4k}{(2\pi)^4} \int \frac{d^4l}{(2\pi)^4} \int \frac{d^4q}{(2\pi)^4} (2\pi)^4 \,\delta\,(p-k-l-q) \times \\ &\times \delta\left(k^2-m^2\right)\theta\,(l_0)\,\delta(l^2-m^2)\theta\,(q_0)\,\delta\left(q^2-m^2\right)\times \\ &\times \varepsilon_\mu\,(p)\,\varepsilon_\nu\,(-p)\,J^{\mu\nu},\end{aligned} \tag{2.86}$$

where

$$J^{\mu\nu} = \frac{1}{8}\left(\frac{ig^2}{2}\right)^4 \frac{(p-q)^4}{m^4}\left(-\eta^{\mu\nu} + \frac{q^\mu q^\nu}{m^2}\right)\left(2 + \frac{(k_\gamma l^\gamma)^2}{m^2}\right). \tag{2.87}$$

The first and the last diagrams are symmetric, and hence provide the same contribution. The imaginary part of these diagrams is given by

$$\begin{aligned}\mathrm{Im}\,(\Gamma_{2B}) = &\int \frac{d^4k}{(2\pi)^4} \int \frac{d^4l}{(2\pi)^4} \int \frac{d^4q}{(2\pi)^4} (2\pi)^4 \,\delta(p-k-l-q)\times \\ &\times \theta\,(k_0)\,\delta\left(k^2-m^2\right)\theta\,(l_0)\,\delta(l^2-m^2)\theta\,(q_0)\,\delta\left(q^2-m^2\right)\times \\ &\times \varepsilon_\mu\,(p)\,\varepsilon_\nu\,(-p)\,I^{\mu\nu},\end{aligned} \tag{2.88}$$

where

$$\begin{aligned}I^{\mu\nu} = &\frac{1}{2}\left(\frac{ig^2}{2}\right)^4 \frac{(p-q)^2}{m^2}\frac{(p-k)^2}{m^2}\times \\ &\times\left(-\eta^{\mu\rho} + \frac{q^\mu q^\rho}{m^2}\right)\left(-\eta^{\xi\nu} + \frac{k^\xi k^\nu}{m^2}\right)\left(-\eta_{\rho\xi} + \frac{l_\rho l_{xi}}{m^2}\right).\end{aligned} \tag{2.89}$$

Collecting the contributions of all diagrams, and using the fact that the momentas k, l, q and p are on-shell, we find that the terms containing the inverse powers of mass do not cancel. Rather, the imaginary part of the 2-loop diagrams with 3-particle cut contributions is given by:

$$\mathrm{Im}\,(\Gamma_2) = \mathrm{Im}\,(\Gamma_{2A}) + \mathrm{Im}\,(\Gamma_{2B})\,, \tag{2.90}$$

where

$$\begin{aligned}\mathrm{Im}\,(\Gamma_2) = &\int \frac{d^4k}{(2\pi)^4}\int \frac{d^4l}{(2\pi)^4}\int \frac{d^4q}{(2\pi)^4}(2\pi)^7\delta(p-k-l-q)\times \\ &\times \theta\,(k_0)\,\delta\left(k^2-m^2\right)\theta\,(l_0)\,\delta(l^2-m^2)\theta\,(q_0)\,\delta\left(q^2-m^2\right)\times \\ &\times \varepsilon_\mu\,(p)\,\varepsilon_\nu\,(-p)\,P^{\mu\nu},\end{aligned} \tag{2.91}$$

and

$$P^{\mu\nu} \sim -\frac{g^8 l^\mu l^\nu}{2^5 m^{10}}\left[(lq)^4 - 4(lq)^2(lk)^2 + (lk)^4\right], \tag{2.92}$$

is the most dominant contribution. We see that this contribution is singular in mass, indicating that the unitarity is violated at two loops. On the first sight, from (2.92), we could conclude that this corresponds to the energy scale $k_d \sim g^{-\frac{4}{5}}m$. This is almost true—the scale k_d signalises the violation of unitarity, but concerns only the corrections to the propagator of the transverse modes. We have seen that the polarisation vectors for the longitudinal modes are also singular in mass. Therefore, for the longitudinal modes on the external lines, we find that the unitarity is violated on the lower scales—$k_u \sim g^{-\frac{2}{3}}m$. This is also in agreement with [105].

2.3 A Change of the Description

So far, we have analyzed the corrections to the propagator of a massive photon using the manifestly covariant description in Sect. 2.1.1.5. We have shown that this theory violates unitarity for energy scales $k_u \sim g^{-\frac{2}{3}}m$. However, we can notice that the manifestly covariant description that we have used so far is not practical for several reasons. The main inconvenience in the form of the propagators (2.56) and (2.65)—for high energies they tend to be a constant and indicate a power counting non-renormalizable theory. Moreover, one of the main questions we wish to answer is if the massless limit of these theories is smooth. While in the case of the toy model this perturbatively does not seem to be the case, as we will see, it is worthwhile to ask the same question in massive Yang-Mills theory. Therefore, the propagators in this form are inconvenient to answer such a question—they are singular in mass.

The Stueckelberg trick—a trick with which one can rewrite the theory with the help of an additional field, *the Stueckelberg field* in such a way that the gauge redundancy is restored—and its non-Abelian generalization [109, 110], were approached to aid the shortcomings of the original manifestly covariant formulation. By using it, the resulting theory remains manifestly covariant, but the propagator becomes power-counting renormalizable.

In this thesis, we will adopt another approach. We have seen that only the longitudinal and transverse modes are the physical degrees of freedom of the theory. Therefore, in order to analyze the massless limit and the high-energy properties of the theory, we will extend the approach we had initially started with in 2.1.1.2 to an interacting theory.

In the chapters to come, we will find that this procedure is non-trivial in the case of the massive Yang-Mills theory. Nevertheless, it will be possible to find it, by generalizing the results of the toy model, that we will now study.

2.3.1 *The Toy Model and the Degrees of Freedom*

Let us begin with the action of the toy model (2.69), and decompose the spatial part of the vector field according to the Helmholz decomposition (2.17) as in the free theory. Then, the action becomes:

$$\begin{aligned} S = \frac{1}{2}\int d^4x \Big\{ & A_0\left(-\Delta + m^2 - \frac{g^2}{2}\Delta\chi\right)A_0 \\ & +2A_0\left[\Delta\dot{\chi} + \frac{g^2}{4}\partial_0\left(A_i^T A_i^T + \chi_{,i}\chi_{,i} + 2A_i^T\chi_{,i}\right)\right] \\ & + \left(\dot{A}_i^T\dot{A}_i^T - A_{i,j}^T A_{i,j}^T - m^2 A_i^T A_i^T\right) - \left(\dot{\chi}\Delta\dot{\chi} - m^2\chi\Delta\chi\right) \\ & + \frac{g^2}{2}\left(A_i^T A_i^T + 2A_i^T\chi_{,i} + \chi_{,i}\chi_{,i}\right)\Delta\chi\Big\}. \end{aligned} \tag{2.93}$$

Here, we have omitted the total derivatives. As in the free case, there are no time derivatives acting on the A_0 component. This means that the temporal part of the vector field is not propagating—it satisfies a constraint that we will now find. By varying the action with respect to A_0 we obtain:

$$\left(-\Delta + m^2 - \frac{g^2}{2}\Delta\chi\right)A^0 = -\Delta\dot{\chi} - \frac{g^2}{4}\partial_0\left(A_i^T A_i^T + \chi_{,i}\chi_{,i} + 2A_i^T\chi_{,i}\right). \tag{2.94}$$

Under the assumption

$$\frac{g^2}{2}\frac{-\Delta}{-\Delta + m^2}[\chi] \ll 1 \tag{2.95}$$

that we will check a posteriori, we can evaluate this constraint as:

$$A_0 = \frac{1}{-\Delta + m^2 - \frac{g^2}{2}\Delta\chi}\left[-\Delta\dot{\chi} - \frac{g^2}{4}\partial_0\left(A_i^T A_i^T + \chi_{,i}\chi_{,i} + 2A_i^T\chi_{,i}\right)\right]. \tag{2.96}$$

By substituting this solution back to the action, we obtain the following Lagrangian density for the energy scales $k^2 \gg m^2$:

$$\mathcal{L} = \mathcal{L}_{0\chi} + \mathcal{L}_{0A^T} + \mathcal{L}_{g\chi} + \mathcal{L}_{g\chi A^T} + \mathcal{L}_{g^2 int}, \qquad \text{where} \tag{2.97}$$

$$\mathcal{L}_{0\chi} = -\frac{m^2}{2}\chi\left(\Box + m^2\right)\frac{-\Delta}{-\Delta + m^2}\chi$$

$$\mathcal{L}_{0A_i^T} = -\frac{1}{2}A_i^T\left(\Box + m^2\right)A_i^T$$

$$\mathcal{L}_{g\chi} \sim \frac{g^2}{2}(\dot{\chi}^2 - \frac{1}{2}\chi_{,i}\chi_{,i})(-\Delta\chi)$$

$$\mathcal{L}_{gA_i^T\chi} \sim -\frac{g^2}{4}(A_i^{T^2} + 2A_i^T\chi_{,i})\Box\chi$$

$$\mathcal{L}_{g^2 int} \sim -\frac{g^4}{8}\left[-\dot{\chi}\Delta\chi + A_i\dot{A}_i\right]\frac{1}{-\Delta}\left[-\dot{\chi}\Delta\chi + A_j\dot{A}_j\right].$$

In order to get the clear insight into the structure of interactions, here we shown only the approximate form of the Lagrangian, valid only for energies $k^2 \gg m^2$. In order to find the Feynman rules, on the other hand, the full Lagrangian is necessary. It is presented in the appendix along with the Feynman rules, up to and including $O(g^4)$.

Similarly to the free theory, the longitudinal modes are not canonically normalised. Normalising them according to (2.33), we obtain:

$$\begin{aligned}
\mathcal{L}_{0\chi} &= \frac{1}{2}\left(\dot{\chi}_n\dot{\chi}_n - \chi_{n,i}\chi_{n,i} - m^2\chi_n\chi_n\right)\\
\mathcal{L}_{0A_i^T} &= \frac{1}{2}\left(\dot{A}_i^T\dot{A}_i^T - A_{i,j}^T A_{i,j}^T - m^2 A_i^T A_i^T\right)\\
\mathcal{L}_{g\chi} &\sim \frac{g^2}{2m^3}\left(\dot{\chi}_n^2 - \frac{1}{2}\chi_{n,i}\chi_{n,i}\right)(-\Delta\chi_n)\\
\mathcal{L}_{gA_i^T\chi} &\sim -\frac{g^2}{4m}\left(A_i^T A_i^T + \frac{2}{m}A_i^T\chi_{n,i}\right)\Box\chi_n\\
\mathcal{L}_{g^2 int} &\sim -\frac{g^4}{8}\left[-\frac{1}{m^2}\dot{\chi}_n\Delta\chi_n + A_i\dot{A}_i\right]\frac{1}{-\Delta}\left[-\frac{1}{m^2}\dot{\chi}_n\Delta\chi_n + A_j\dot{A}_j\right].
\end{aligned} \tag{2.98}$$

We can right away see several benefits such formulation would bring us. In contrast to theory *a la Stueckelberg*, this approach has no gauge redundancy—everything is presented directly through the degrees of freedom of the theory. Furthermore, the propagators (2.46) and (2.47) are now power counting renormalizable. However, similarly to the Stueckelberg trick, the singularities in mass are now appearing in the interactions. Nonetheless, we can see that this approach will be more suitable to study the massless limit, as opposed to the original manifestly covariant formulation—the propagator of the transverse modes in the massless limit matches with the propagator of the transverse modes in massless theory in the Coulomb gauge. The price we had to pay, on the other hand, is the loss of the manifest Lorentz covariance.

One of the biggest advantages of this formulation is that it keeps the degrees of freedom explicit. This will also hold in the case of diagrams. However, the Feynman rules in this formulation take a particularly aesthetically unpleasing form if one works with the canonically normalized modes. With a bit of care, one can always work with the original modes and rescale the external lines such that the formulation in terms of the normalized modes is obtained at the end of the calculation.

2.3.2 *From Manifestly Covariant to Manifestly Non-covariant Diagrams*

Let us now see how to connect the diagrams of the manifestly covariant formulation with the one that includes only the transverse and longitudinal modes. For simplicity, in the following, we will use the original longitudinal modes instead of the canonically normalized ones. The results will be equivalent for both choices, and one can easily go from one to another one with the use of (2.33) (Fig. 2.11).

In order to find the rules that govern the translation of a manifestly covariant diagram into a manifestly non-covariant one we first have to think to which propagator of manifeslty non-covariant theory does the manifestly covariant propagator correspond to—either to the propagator of transverse or of the longitudinal mode. In the momentum space, we have seen that these are given by

$$\begin{aligned} \Delta^T_{ij}(k) &= \left(\delta_{ij} - \frac{k_i k_j}{\boldsymbol{k}^2}\right)\frac{i}{k^2 - m^2} \qquad \text{for transverse modes} \\ \Delta_\chi(k) &= \frac{1}{m^2 A(k)}\frac{i}{k^2 - m^2} \qquad \text{for the (original) longitudinal modes,} \end{aligned} \tag{2.99}$$

where

$$A(k) = \frac{\boldsymbol{k}^2}{\boldsymbol{k}^2 + m^2}. \tag{2.100}$$

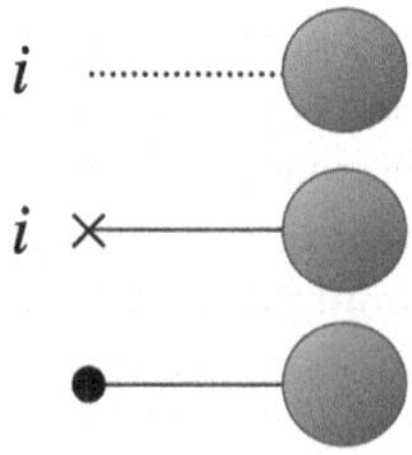

Fig. 2.11 The type of the propagator lines. Reprinted under CC-BY-4.0 license from [REF_NR]. ©2022, The Author(s)

If the propagator line is an external one and corresponds to a longitudinal mode, then there are two possibilities that should multiply it, as depicted in the picture. There, the grey part denotes the remaining part of the diagram. With a cross, it comes with a factor $-ip_i$, while for a circle it comes with $-ip_0A(p)$. The reason for this can be traced back to the manifestly-covariant propagator. There, the longitudinal modes appear both in the purely spatial part of the propagator, which corresponds to the first factor, and the temporal part of the propagator—in place of the A_0 component. The rules to transform a diagram in a manifestly covariant theory to the ones that contain only the physical degrees of freedom are given by the following [66]:

1. Find all external propagator combinations.
2. Connect them with every possible vertex.
3. Write the expression for the diagram by using Feynman rules for the manifestly covariant theory.[2]
4. Multiply the diagram with factors from three propagator types.

As an example, let's consider the 1-loop correction to the propagator. Then all possible diagrams one could form are given by the following picture (Fig. 2.12):

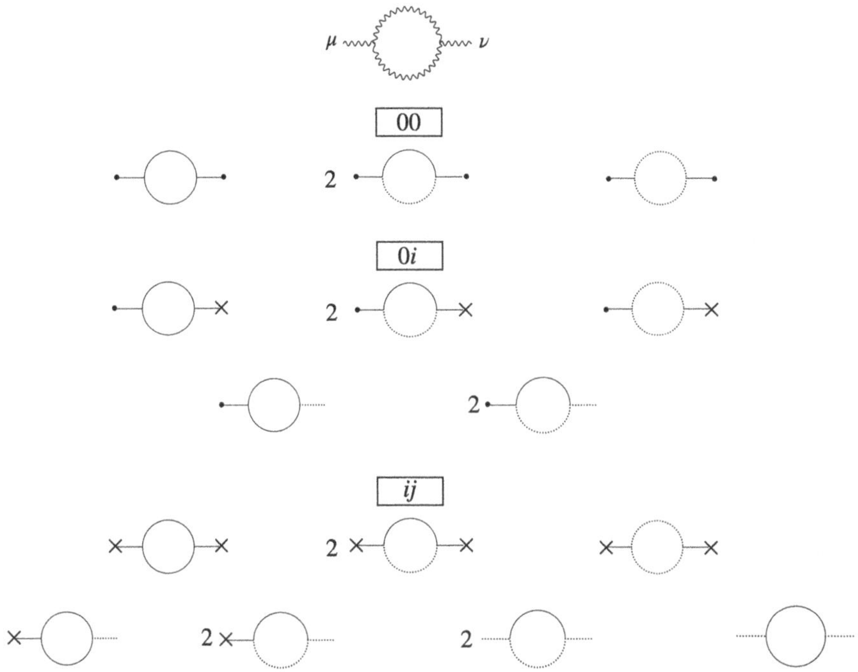

Fig. 2.12 The one-loop corrections from a manifestly covariant to a manifestly non-covariant theory. Reprinted under CC-BY-4.0 license from [REF_NR]. ©2022, The Author(s)

[2] In the case of the toy model, they are given in the appendix.

Here, the dotted line corresponds to the transverse modes, while the full one corresponds to the longitudinal modes.

In comparison to the manifestly covariant formulation, it is now very apparent where the divergences come from, and how the longitudinal and transverse modes appear in the diagrams. Moreover, let us demonstrate the unitarity violation of the transverse modes at two loops, given by the following diagrams (Figs. 2.13, 2.14, and 2.15):

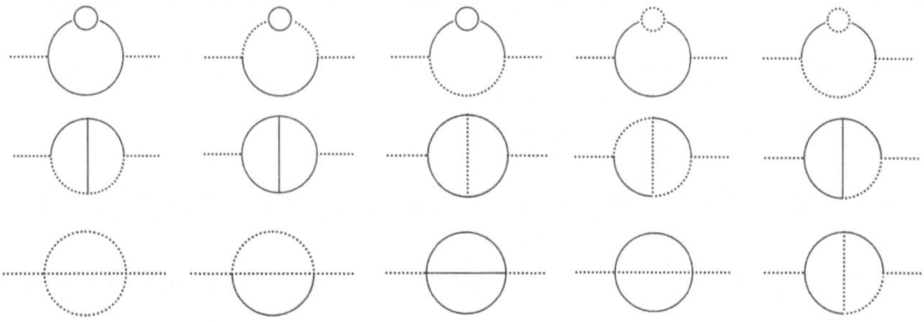

Fig. 2.13 Diagrams responsible for unitarity violation of the transverse modes at two-loops

In order to estimate the contribution of each particular diagram, all we have to do is count in a factor of $\frac{1}{m^2}$ for each longitudinal propagator—these are the ones that are cause a discontinuity. Therefore, on the one hand, we can see that a diagram that has no divergences in the massless limit contains purely transverse modes:

Fig. 2.14 The least divergent diagram at two-loops in the toy model

The contribution of this diagram to the corrections to the propagator of the transverse modes can be estimated as $\sim g^8 k^2$. On the other hand, the most divergent diagrams are ones that contain only the longitudinal modes in the inner lines:

Fig. 2.15 The most divergent diagram at two-loops in the toy model

These can be estimated as $\sim \frac{g^8 k^{12}}{m^{10}}$, thus indicating that the unitarity for the transverse modes is violated for the scales $k_d \sim g^{-\frac{4}{5}} m$, in agreement with the result that we have previously found in the manifestly covariant analysis.

2.4 The Puzzle of Massive Yang-Mills Theory

2.4.1 *Renormalizability and Unitarity*

In the previous sections, we have seen that the standard diagrammatic approach suggests that the toy model does not have a smooth massless limit—the two-loop corrections to the propagator are singular with mass, leading to the apparent violation of unitarity. If we would replace this self-interaction with a conserved source, the resulting theory would be renormalizable and unitary [45, 111]—the singular part of the propagator would drop out of the equations. Yet, the research over the past sixty years has shown that this does not seem to be the case for the massive Yang-Mills theory, even if the source is conserved.

Similarly to the Proca theory, the propagator of the massive Yang-Mills theory is singular in mass. As a result, it tends to a constant at high energies, which suggests that the theory is not power-counting renormalizable. Yet, due to the results of Proca theory coupled with a conserved source such an argument could be misleading. The first attempts that studied the renormalizability of massive Yang-Mills theory have thus rewritten the theory *a la Stueckelberg* [46–49, 51, 53]. Even though the propagators in this formulation are power-counting renormalizable, the authors have nevertheless found that the singularities in mass cannot be avoided and thus concluded that the theory is not renormalizable.

Let us look at this result from another perspective [46]. By setting the external source to zero, the equations of motion of massive Yang-Mills theory are given by

$$D_\mu F^{\mu\nu a} + m^2 A^{\nu a} = 0. \tag{2.101}$$

We can rewrite these equations in the following way:

$$f^{\mu\nu a}_{,\mu} + m^2 A^{\nu a} = j^{\nu a}, \tag{2.102}$$

where

$$f^{\mu\nu a} = A^{\mu,\nu a} - A^{\nu,\mu a}, \tag{2.103}$$

and

$$j^{\nu a} = g\varepsilon^{abc}\left[\partial_\mu\left(A^{\mu b}A^{\nu c}\right) + A^b_\mu F^{\mu\nu c}\right]. \tag{2.104}$$

These equations remind us of equations for three Proca fields, where the source for each one of $a = 1, 2, 3$ is conserved:

$$j^{\nu a}_{,\nu} = 0. \tag{2.105}$$

Therefore, the above result is very surprising—a single Proca theory with a conserved source is renormalizable, while three of them with conserved sources for each, albeit nonetheless with coupled equations of motion are not.

In comparison with the toy model, the divergences in the massive Yang-Mills theory are smaller despite the presence of a cubic self-interaction in both theories. As pointed out in [46], only one Stueckelberg field decouples from the remaining degrees of freedom, reducing the overall singular behavior of the scattering matrix. The decoupling of the remaining ones was not apparent.

The first attempts [46–49] were criticized by Vainshtein and Khriplovich [58] and Slavnov [52] for missing some of the interactions after applying the Stueckelberg decomposition and using the wrong form of the propagators. However, this did not present a major issue at high energies. Their conclusion has agreed with later studies of [51], where functional integral techniques were applied, and [53], whose aim was to construct generalized Ward identities in order to analyze the divergences. These methods have led to the same conclusion—at one loop the theory is finite, but once we reach two loops and more, the theory does not seem to be renormalizable.

Renormalisability in the sense of an effective field theory—a perspective in which the divergences are removed from physical quantities with a redefinition of an infinite number of parameters in the effective Lagrangian—was recently argued as a possibility for massive Yang-Mills theory [112]. Even so, this was not the only issue that the theory had—the perturbative methods suggested that it also violated unitarity.

The development of cutting rules has initiated the tests of unitarity in massive Yang-Mills theory [50, 54–58]. There were two interesting cases—one with only the transverse modes on the external lines that were motivated by the absence of the longitudinal modes in the massless theory, and one where the longitudinal modes were also taken into account.

The first case has revealed that the theory was unitary at one loop [50]. However, similar to the case of renormalizability, this was no longer the case for higher-order corrections, starting with two loops [54]. The second case, on the other hand, has shown a violation of unitarity even at the tree level [55–58]. The amplitude was proportional to the squared ratio of the coupling constant and the mass of the vector field, indicating that the unitarity is violated at length-scales

$$L_u \sim \frac{g}{m}. \tag{2.106}$$

At higher loops, it became apparent in [53] that the divergences appear in the form of the perturbative series, based on powers of $\frac{g^2\Lambda^2}{m^2}$, where Λ is the energy-cutoff of the theory. The same series was also found in [50], by analyzing the n-point functions that involved only the longitudinal modes. This brought an opportunity—*Could the divergences be resumed and thereafter allow for a smooth massless limit?*

Although this possibility seemed promising, it was soon excluded by a surprising result. The finite corrections to the propagator of the transverse modes at one loop are finite. However, they do not match with those of massless theory once the mass is taken to zero [59, 113, 114]. Let us demonstrate this result in more detail by rewriting the theory in terms of the longitudinal and transverse modes—the physical degrees of freedom.

2.4.2 The Discontinuity and the Physical Degrees of Freedom

Following the approach described in the toy model, we will start by rewriting the action in terms of a temporal and spatial part of the vector field. Then, the Lagrangian density corresponding to the action (2.57) becomes

$$\begin{aligned}\mathcal{L} =& \frac{1}{2}\left[A_0^a\left(-\Delta+m^2\right)A_0^a + 2A_0^a\Delta\dot{A}_{i,i}\right]\\ &\frac{1}{2}\left(+\dot{A}_i^a\dot{A}_i^a + A_{i,i}^a A_{j,j}^a - A_{i,j}^a A_{i,j}^a - m^2 A_i^a A_i^a\right)\\ &+ g\varepsilon^{abc}\left[A_0^a\dot{A}_i^b A_i^c - A_{0,i}^a A_i^b A_0^c + A_{j,i}^a A_i^b A_j^c\right]\\ &\frac{1}{2}g^2\varepsilon^{abc}\varepsilon^{ade}\left(A_0^b A_i^c A_0^d A_i^e - \frac{1}{2}A_i^b A_j^c A_i^d A_j^e\right).\end{aligned} \tag{2.107}$$

We can right away notice a similarity with the Proca theory—the temporal part of the vector field is not propagating. However, instead of having one constraint we will now have three of them, one for each $a = 1, 2, 3$. By varying the action (2.107) with respect to A_0^a, we obtain

$$\begin{aligned}\left(-\Delta+m^2\right)A_0^a =& -\dot{A}_{i,i}^a - g\varepsilon^{abc}\dot{A}_i^b A_i^c - g\varepsilon^{abc}A_i^b A_{0,i}^c - g\varepsilon^{abc}\partial_i\left(A_i^b A_0^c\right)\\ &- g^2\epsilon^{fbc}\varepsilon^{fad}A_0^b A_i^c A_i^d.\end{aligned} \tag{2.108}$$

Up to $O\left(g^2\right)$, its solution can be evaluated as

$$\begin{aligned}A_0^a =& K\left[\dot{A}_{i,i}^a\right] - \frac{g\varepsilon^{abc}}{-\Delta+m^2}\left[\dot{A}_i^b A_i^c + (A_{i,i}^b + 2A_i^b\partial_i)K\left[\dot{A}_{j,j}^c\right]\right]\\ &+ \frac{g^2\varepsilon^{fab}\varepsilon^{fcd}}{-\Delta+m^2}\left\{A_i^b A_i^c K\left[\dot{A}_{j,j}^d\right]\right.\\ &\left.+(A_{i,i}^b + 2A_i^b\partial_i)\frac{1}{-\Delta+m^2}\left[\dot{A}_j^c A_j^d + (A_{j,j}^c + 2A_j^c\partial_j)K\left[\dot{A}_{k,k}^d\right]\right]\right\},\end{aligned} \tag{2.109}$$

where

$$K[\chi] = \frac{-1}{-\Delta+m^2}\chi. \tag{2.110}$$

We will now insert this constraint back into the action, to express it solely in terms of the physical degrees of freedom. In addition, motivated by the similarity with Proca theory (2.102) let's decompose the spatial part of the vector field as follows:

$$A_i^a = A_i^{Ta} + \chi_{,i}^a, \qquad \text{where} \qquad A_{i,i} = 0. \tag{2.111}$$

For scales $k^2 \sim \frac{1}{L^2} \sim m^2$, the Lagrangian density with the most relevant terms is then given by

$$\mathcal{L} = \mathcal{L}_0 + \mathcal{L}_{int}, \qquad \text{where} \tag{2.112}$$

$$\mathcal{L}_0 = -\frac{1}{2}\chi^a\left(\Box + m^2\right)\frac{m^2(-\Delta)}{-\Delta + m^2}\chi^a - \frac{1}{2}A_i^{Ta}\left(\Box + m^2\right)A_i^{Ta} \qquad \text{and}$$

$$\begin{aligned}\mathcal{L}_{int} \sim g\varepsilon^{abc}&\left[\frac{1}{2}\chi^b\chi_{,i}^c\Box A_i^{Ta} - \dot{\chi}^a\chi_{,i}^c\frac{m^2}{\Delta}\left(\dot{\chi}_{,i}^b\right) + \chi^b A_i^{Tc}\Box A_i^{Ta}\right]\\ &+ \frac{g^2}{2}\varepsilon^{fab}\varepsilon^{fcd}\left[\dot{\chi}^a\dot{\chi}^c\chi_{,i}^b\chi_{,i}^d - \frac{1}{2}\chi_{,i}^a\chi_{,i}^c\chi_{,j}^b\chi_{,j}^d\right.\\ &\left.+\left(\dot{\chi}_{,i}^a\chi_{,i}^b + \dot{\chi}^a\Delta\chi^b\right)\frac{1}{\Delta}\left(\dot{\chi}_{,j}^c\chi_{,j}^d + \dot{\chi}^c\Delta\chi^d\right)\right].\end{aligned}$$

The $\mathcal{L}_0$ corresponds to the kinetic part for the longitudinal and transverse modes. We can see that it has the same form as in Proca theory, differing just in the number of the longitudinal and transverse modes. Moreover, we can notice that the longitudinal modes are again not canonically normalised. Substituting

$$\chi_n^a = m\sqrt{\frac{-\Delta}{-\Delta + m^2}}\chi^a, \tag{2.113}$$

where χ_n are the normalised modes, we obtain

$$\mathcal{L} = \mathcal{L}_0 + \mathcal{L}_{int}, \qquad \text{where} \tag{2.114}$$

$$\mathcal{L}_0 = -\frac{1}{2}\chi_n^a\left(\Box + m^2\right)\chi_n^a - \frac{1}{2}A_i^{Ta}\left(\Box + m^2\right)A_i^{Ta} \qquad \text{and}$$

$$\begin{aligned}\mathcal{L}_{int} \sim g\varepsilon^{abc}&\left[\frac{1}{2m^2}\chi_n^b\chi_{n,i}^c\Box A_i^{Ta} - \frac{1}{m}\dot{\chi}_n^a\chi_{n,i}^c\frac{1}{\Delta}\left(\dot{\chi}_{n,i}^b\right) + \frac{1}{m}\chi_n^b A_i^{Tc}\Box A_i^{Ta}\right]\\ &+ \frac{g^2}{2m^4}\varepsilon^{fab}\varepsilon^{fcd}\left[\dot{\chi}_n^a\dot{\chi}_n^c\chi_{n,i}^b\chi_{n,i}^d - \frac{1}{2}\chi_{n,i}^a\chi_{n,i}^c\chi_{n,j}^b\chi_{n,j}^d\right.\\ &\left.+\left(\dot{\chi}_{n,i}^a\chi_{n,i}^b + \dot{\chi}_n^a\Delta\chi_n^b\right)\frac{1}{\Delta}\left(\dot{\chi}_{n,j}^c\chi_{n,j}^d + \dot{\chi}_n^c\Delta\chi_n^d\right)\right].\end{aligned}$$

Let us notice the similarity with the Stueckelberg trick. It was performed in order to replace the singular propagator with those that indicate a power counting renormalizable theory. This procedure also allows us to do that, without any need of introducing additional degrees of freedom. The propagators of the normalised longitudinal and transverse modes in the momentum space are now given respectively by

$$\Delta^{ab}_{\chi_n}(k) = \frac{i\delta^{ab}}{k^2 - m^2}, \qquad \text{and} \qquad \Delta^{Tab}_{ij}(k) = \left(\delta_{ij} - \frac{k_i k_j}{|\boldsymbol{k}|^2}\right) \frac{i\delta^{ab}}{k^2 - m^2}. \tag{2.115}$$

These are now power-counting renormalizable. However, the singularities in mass have not disappeared—they are now present in the interactions, given by $\mathcal{L}_{int}$. We can also notice that the divergences will be smaller than in the toy model. At $O(g)$, the most divergent term is an interaction of longitudinal and transverse modes. The term that contains only the longitudinal modes is only inversely proportional to the mass, instead of its cubic power as was the case for the toy model. This is due to the additional symmetry of the theory.

Let us now study the imaginary part of the 1-loop corrections to the propagator of the transverse modes. Our goal is to compare these to the ones of the massless theory and show that in the limit $m \to 0$, the results of the two theories do not match. As a first step, we will therefore need Feynman's rules. These, together with the details of the massless theory are given in the appendix. At 1-loop, the corrections to the propagator of the transverse modes in the massive case are given by (Figs. 2.16 and 2.17)

Fig. 2.16 The one-loop corrections to the propagator of transverse modes in massive Yang-Mills theory. Reprinted under CC-BY-4.0 license from [REF_NR]. ©2022, The Author(s)

while in the massless case only the first and fourth diagrams are present. Here, the dotted line represents the propagator for the transverse modes, while the full line corresponds to the longitudinal modes. As we have seen in the previous parts of this thesis, the imaginary part of the diagrams corresponds to the sum of all possible cuts. Therefore, the last two diagrams can be set to zero—they will not contribute to the imaginary part. The first three diagrams, on the other hand, contribute, with cuts given by:

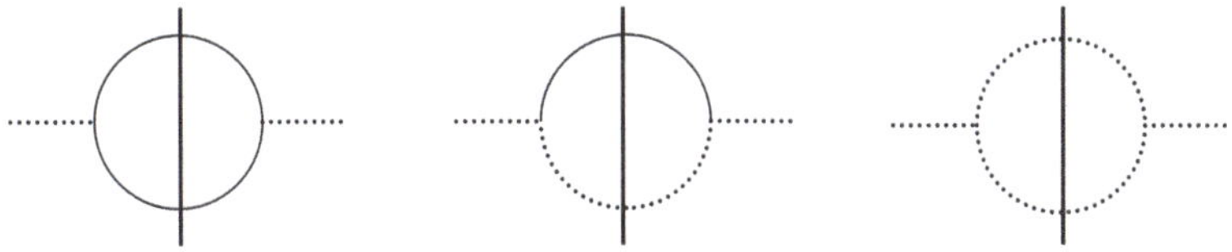

Fig. 2.17 The cuts of the one-loop corrections to the propagator of transverse modes in massive Yang-Mills theory. Reprinted under CC-BY-4.0 license from [REF_NR]. ©2022, The Author(s)

Due to the similarity with the toy model, the Cutkosky rules are easily generalised—a cut of the propagator corresponds to replacing it with the following expression

$$\tilde{\Delta}_{\chi}^{ab}(k) = 2\pi\theta\left(k_0\right)\delta\left(k^2 - m^2\right)\frac{|\boldsymbol{k}|^2 + m^2}{m^2|\boldsymbol{k}|^2}, \qquad \text{for the longitudinal modes, and}$$

$$\tilde{\Delta}_{ij}^{Tab}(k) = 2\pi\theta\left(k_0\right)\delta\left(k^2 - m^2\right)\delta^{ab}\left(\delta_{ij} - \frac{k_i k_j}{|\boldsymbol{k}|^2}\right) \qquad \text{for the transverse modes.}$$

The massless theory is evaluated in the radiation gauge. Cutting the propagator means to set it's momenta on-shell by making the following replacement:

$$\tilde{\Delta}_{ij}^{Tab}(k) = 2\pi\theta\left(k_0\right)\delta\left(k^2\right)\delta^{ab}\left(\delta_{ij} - \frac{k_i k_j}{|\boldsymbol{k}|^2}\right).$$

In each of these diagrams we have to take into account polarisation vectors $\varepsilon_{i\sigma}^{a}(p)$ due to the transverse modes, that are on-shell, $p_i \varepsilon_i^a(p) = 0$, and satisfy the following completeness relation:

$$\sum_{\sigma=1,2}\varepsilon_{i\sigma}^{a}(p)\,\varepsilon_{j\sigma}^{b}(p) = \left(\delta_{ij} - \frac{p_i p_j}{|\boldsymbol{p}|^2}\right)\delta^{ab}.$$

Let us now evaluate the diagrams. The first one contains only the transverse modes. Its imaginary part is given by:

$$\begin{aligned}\mathrm{Im}\Gamma^{T} &= \int\frac{d^4k}{(2\pi)^3}\int\frac{d^4q}{(2\pi)^3}(2\pi)^4\,\delta^{(4)}(p-k-q)\times \\ &\times\theta\left(k_0\right)\delta\left(k^2 - m^2\right)\theta\left(q_0\right)\delta\left(q^2 - m^2\right)\varepsilon_i^a(p)\,\varepsilon_j^b(-p)\,T_{ij}^{ab},\end{aligned} \tag{2.116}$$

where

$$T_{ij}^{ab} = V_{ikl}^{acd}(-p,k,q)\,V_{jnz}^{bcd}(p,-k,-q)\left(\delta_{kn} - \frac{k_k k_n}{|\boldsymbol{k}|^2}\right)\left(\delta_{lz} - \frac{q_l q_z}{|\boldsymbol{q}|^2}\right), \tag{2.117}$$

for the massive case, and

$$\begin{aligned}\mathrm{Im}\Gamma^{T} &= \int\frac{d^4k}{(2\pi)^3}\int\frac{d^4q}{(2\pi)^3}(2\pi)^4\,\delta^{(4)}(p-k-q)\times \\ &\times\theta\left(k_0\right)\delta\left(k^2\right)\theta\left(q_0\right)\delta\left(q^2\right)\varepsilon_i^a(p)\,\varepsilon_j^b(-p)\,T_{ij}^{ab},\end{aligned} \tag{2.118}$$

for the massless fields. T^{ab}_{ij} coincides in both cases, along with the vertex. As these diagrams contain only the transverse modes, there are no divergences and they exactly cancel in the massless limit.

Let us now evaluate the last diagram. Both internal moments are put on shell due to the cut. Therefore, both of the vertices take the following form:

$$V^{cad}_{il} = -igm^2 \varepsilon^{cad} \delta_{il} \frac{1}{|\boldsymbol{q}|^2 + m^2} \left(2q_n k_n + m^2 |\boldsymbol{q}|^2 \right). \tag{2.119}$$

The first one cancels with the singular part of the propagator for the longitudinal modes, and therefore, the imaginary part of this diagram is overall multiplied by m^2. As a result, its contribution vanishes in the massless limit.

The final diagram that is left to evaluate is the middle one. It's imaginary part is given by

$$\begin{aligned} \mathrm{Im}\Gamma_L = & \int \frac{d^4k}{(2\pi)^3} \int \frac{d^4q}{(2\pi)^3} (2\pi)^4 \delta^{(4)}(p-k-q) \times \\ & \times \theta(k_0)\, \delta\left(k^2 - m^2\right) \theta(q_0)\, \delta\left(q^2 - m^2\right) \varepsilon^a_i(p)\, \varepsilon^b_j(-p)\, I^{ab}_{ij}, \end{aligned} \tag{2.120}$$

where

$$I^{ab}_{ij} = -\frac{1}{4} \delta^{ce} \delta^{df} \frac{|\boldsymbol{k}|^2 + m^2}{|\boldsymbol{k}|^2} \frac{|\boldsymbol{q}|^2 + m^2}{|\boldsymbol{q}|^2} V^{acd}_{i,2\chi}(-p, k, q)\, V^{bef}_{j,2\chi}(p, -k, -q). \tag{2.121}$$

For $k^2 = q^2 = m^2$ we have

$$V^{acd}_{i,2\chi}(-p, k, q) = -gm^2 \varepsilon^{acd} k_i \left[1 + \frac{2m^2 k_0 q_0}{(|\boldsymbol{k}|^2 + m^2)(|\boldsymbol{q}|^2 + m^2)} \right], \tag{2.122}$$

and similar expression for the other vertex. Keeping only the terms that survive in the massless limit, we therefore obtain

$$I^{ab}_{ij} = \frac{1}{2} g^2 \delta^{ab} k_i k_j. \tag{2.123}$$

This contribution is the cause of the discontinuity with the massless theory, with a factor that agrees with the result of [59]. It is purely due to the longitudinal modes, that fail to decouple from the transverse modes.

This behavior is puzzling from the point of view of the physical continuity [44]. It implies that an experiment should not be able to distinguish between a massless particle and a particle with an infinitesimally small mass [115]. So far, it seems that massive Yang-Mills theory does not satisfy this principle. But, is this argument sufficient to conclude that the massless limit of the massive Yang-Mills theory is not smooth?

2.4.3 *The Choice of a Decomposition*

So far, we have considered perturbative aspects of the massive Yang-Mills theory. The standard perturbative methods suggest that the massless limit of this theory is not smooth. Nevertheless, soon after the discovery of the perturbative discontinuity of this theory A. I. Vainshtein and I. B. Khriplovich have suggested that this discontinuity could resolved beyond the perturbation theory in [58]—*"...it appears highly probable that outside perturbation theory, a continuous zero-mass limit exists and the theory is renormalizable."*. Similar to the previous approaches, they have used a Stueckelberg decomposition. However, it was a non-linear one, initially suggested in [110]. As a result, the Lagrangian became non-polynomial in fields. The authors of [58] have conjectured that if these terms were not expanded, the resulting theory would renormalizable and have a smooth massless limit. However, [52] pointed out that the proof to this conjecture is still absent—one had to show that the massless limit is smooth not only at the level of Lagrangian, but that the same also holds for the matrix elements. This has led to subsequent studies that focused on finding a way to analyze Lagrangian that contained non-polynomial terms.

In a promising attempt, [116–118] have derived conditions for unitarity in the Landau gauge theory and proposed a strategy to subtract the divergences. These criteria, however, cannot be applied to the unitary gauge. Moreover, in [119] it was shown that the non-polynomial theory can be algebraically reduced to a polynomial form by applying several field redefinitions that do not alter the S-matrix. The final polynomial form has resembled the one of [49], implying that the theory is not renormalizable and unitary [120]. Yet, in order to understand if this conclusion, which is in agreement with most of the previous studies is indeed sufficient to conclude that the massless limit of massive Yang-Mills theory is not smooth, we must look at what changes if one performs a field redefinition.

By construction, a field redefinition of [119] does not change the S-matrix. However, it changes the definition of the degrees of freedom. On the one hand, the Lagrangian polynomial in fields can be obtained through a linear decomposition of a vector field. This decomposition was used in [46–49], and corresponds to the linear relation between field U^a_μ of the original theory and the one in the theory *a la Stueckelberg*:

$$U^a_\mu = A^a_\mu + \frac{1}{m} B^a_{,\mu}, \tag{2.124}$$

where B^a are the Stueckelberg fields, and A^a_μ new vector fields. The non-linear decomposition, on the other hand, introduces non-polynomial terms in the Lagrangian. The relation between a vector field of the original field and the new vector and Stueckelberg field, in this case would be non-linear [110]:

$$U_\mu = S^{-1} A_\mu S + \frac{i}{g} S^{-1} S_{,\mu}, \tag{2.125}$$

where S is a unitary matrix with longitudinal modes as its parameters. Thus the following is unclear—*Which decomposition to use?*

The purpose of this work is to show that discontinuity that appears in the massive Yang-Mills theory is just an artifact of the standard perturbation theory. This is due to the existence of an underlying mechanism—the Vainshtein mechanism—that was originally discovered in the context of massive gravity [63]. In the following chapters, we will study this mechanism in detail and apply it to the theories of vector fields that we have thus far considered. With methods that originate from the cosmological perturbation theory, we will show that not only is the non-linear decomposition the only correct one, but that due to the self-interacting terms, the longitudinal modes become strongly coupled at the Vainshtein scale, beyond which they decouple from the transverse modes up to small corrections. This will show that the massless limit is smooth, as initially proposed in [58].

Chapter 3
The Vainshtein Mechanism

The standard methods of quantum field theory have so far served us to explore the perturbative properties of massive vector field theories. These methods indicate that the predictions of the massive Yang-Mills theory differ from those of the massless theory. As we have seen, the imaginary part of the one-loop corrections to the propagator of the transverse modes differs from that of the massless theory by a factor of $\frac{1}{2}$ even in the massless limit. Moreover, when evaluated at higher loops, the perturbative series are singular in mass, thus indicating that the theory violates unitarity.

From the perspective of the physical continuity, the presence of the perturbative discontinuity is curious – we would expect that a small modification, such as the addition of a small mass term would not affect our predictions to such an extent. Yet, this puzzle is not only present in massive vector theories.

Linearised massive gravity – a theory of a massive spin-2 field with a mass term of the Fierz-Pauli form [60] – has been shown to suffer from an even more surprising discontinuity. Its predictions differ from those of General Relativity not only in one-loop corrections, as was the case for massive Yang-Mills theory, but even for three-level diagrams [59, 61, 62]. The reason for this discrepancy is the longitudinal mode – a degree of freedom of massive gravity that is absent in Einstein's gravity. Similar to massive vector theories that we have studied so far, it fails to decouple from the remaining degrees of freedom. Yet, if one would exclude massive gravity as a possible theory of nature, based on the vDVZ discontinuity, one would soon come to regret it. Soon after the discovery of this discontinuity came a resolution to the puzzle – the Vainshtein mechanism [63, 64].

The core of this mechanism lies in the fact that massive gravity is a non-linear theory. The perturbative approach is thus only valid as long as the non-linear terms are smaller than the linear ones in the equations of motion. However, as pointed out by A. Vainshtein, at a certain scale – the Vainshtein radius – this will no longer be satisfied. As a result, the longitudinal mode enters a strong coupling regime due to the non-linear terms, where it remains for all scales smaller than the Vainshtein

A. Hell, *The Massless Limit of Massive Gauge Theories*, Springer Theses,
https://doi.org/10.1007/978-3-032-01090-2_3

radius. Thus for length scales smaller than the Vainshtein radius, the agreement between the predictions of massive gravity and general relativity is restored up to small corrections [65].

One of the central questions that we wish to answer in this thesis is the following: *Could the Vainshtein mechanism also be the underlying mechanism of massive Yang-Mills theory and allow for a smooth massless limit?* In order to explore such a possibility, in this chapter, we will first study the key features of this mechanism in massive gravity. Then, we will apply its ideas to the toy model and return to the massive Yang-Mills theory in the next chapter.

3.1 The Steps to a Smooth Massless Limit

The action of linearised massive gravity – known as the Fierz-Pauli action – given by the following expression:

$$\begin{aligned} S = \frac{1}{8} \int d^4x \, \Big[& h^{\mu\nu,\alpha} h_{\mu\nu,\alpha} - 2h^{\mu\alpha}_{\ \ ,\mu} h^{\nu}_{\alpha,\nu} + 2h_{,\mu} h^{\nu\mu}_{\ \ ,\mu} - h_{,\mu} h^{,\mu} \\ & + m^2 \left(h^2 - h^{\mu\nu} h_{\mu\nu} \right) \Big] \end{aligned} \tag{3.1}$$

where $h_{\mu\nu}$ is the tensor field that describes the massive graviton, $h = \eta^{\mu\nu} h_{\mu\nu}$ is its trace and m is the mass of the graviton. Throughout this part, we will raise and lower the indices with the Minkowski metric $\eta_{\mu\nu}$. In addition, we have set $8\pi G = 1$ in the above action. In order to understand its content, let us first study the physical degrees of freedom of this theory. This will be done following [121]. There, the graviton was made massive via the Higgs mechanism. Nevertheless, the resulting linearised part of the obtained action coincides with (3.1) in the unitary gauge.

3.1.1 The Degrees of Freedom of Massive Gravity

Following the approach of [121, 122], we will decompose $h_{\mu\nu}$ similarly to decomposition in the cosmological perturbation theory [38, 123], according to the irreducible representations of the SO(3) group:

$$\begin{aligned} h_{00} &= 2\phi \\ h_{0i} &= S_i + B_{,i} \\ h_{ij} &= 2\psi\delta_{ij} + 2E_{,ij} + F_{i,j} + F_{j,i} + h^T_{ij} \end{aligned} \tag{3.2}$$

where

$$S_{i,i} = 0 \qquad F_{i,i} = 0 \qquad h^T_{ij,j} = 0 \qquad h^T_{ii} = 0 \tag{3.3}$$

Then, we can separate action into three cases according to the type of modes – the scalar, vector and tensor perturbations.

3.1.1.1 The Scalar Modes

The action for the scalar perturbations is given by

$$S_S = \int d^4x \Big[-3\dot{\psi}\dot{\psi} - \psi\Delta\psi + 3m^2\psi^2 + 2\psi\Delta\left(\ddot{E} - \dot{B}\right) \\ +\phi\left(2\Delta\psi - 3m^2\psi - m^2\Delta E\right) + m^2\left(2\psi\Delta E - \frac{1}{4}B\Delta B\right)\Big] \tag{3.4}$$

Among the types of perturbations, this action is the most complicated one. Even though we have four scalar potentials, only one of them is describing a scalar degree of freedom. Let us find it, by searching for the constraints satisfied by the remaining ones, resolving them and substitute back to the (3.4). First, we can notice that ϕ is a Lagrange multiplier. By varying (3.4) with respect to it, we obtain the following constraint:

$$2\Delta\psi - m^2\left(3\psi + \Delta E\right) = 0, \tag{3.5}$$

that we will solve for E:

$$E = \left(\frac{2}{m^2} - \frac{3}{\Delta}\right)\psi. \tag{3.6}$$

Substituting this back into (3.4), we obtain

$$S_S = \int d^4x \left(-\frac{4}{m^2}\dot{\psi}\Delta\dot{\psi} + 3\dot{\psi}\dot{\psi} + 3\psi\Delta\psi - 3m^2\psi^2 - \frac{m^2}{4}B\Delta B + 2B\Delta\dot{\psi}\right). \tag{3.7}$$

Varying this expression with respect to B, we find another constraint

$$-\frac{m^2}{2}\Delta B + 2\Delta\dot{\psi} = 0, \tag{3.8}$$

whose solution is given by

$$B = \frac{4}{m^2}\dot{\psi}. \tag{3.9}$$

Substituting this into (3.7), we obtain

$$S_S = \int d^4x \left(3\dot{\psi}\dot{\psi} + 3\psi\Delta\psi - 3m^2\psi^2\right) \tag{3.10}$$

This action describes the scalar mode of the massive graviton. As we will see, this mode is the reason for the appearance of the vDVZ pathology, once we consider the perturbations in the presence of external matter.

3.1.1.2 The Vector Modes

The action for the vector modes is given by the following expression:

$$S_V = \frac{1}{4}\int d^4x \left[S_i\left(-\Delta + m^2\right)S_i + 2S_i\Delta\dot{F}_i - \dot{F}_i\Delta\dot{F}_i + m^2F_i\Delta F_i\right] \tag{3.11}$$

we can notice that there are two types of vector modes – F_i and S_i. However, S_i is not propagating. Similarly to the A_0 component in Proca theory, no time derivatives act on it. By varying the action with respect to S_i, we obtain the following constraint:

$$\left(-\Delta + m^2\right)S_i = -\Delta\dot{F}_i, \tag{3.12}$$

that we can resolve as

$$S_i = \frac{-\Delta}{-\Delta + m^2}\dot{F}_i \tag{3.13}$$

Substituting this back into (3.11), we obtain

$$S_V = -\frac{1}{4}\int d^4x F_i\left(\Box + m^2\right)\frac{-\Delta m^2}{-\Delta + m^2}F_i. \tag{3.14}$$

Thus, the above action describes two vector degrees of freedom. In general relativity, these degrees of freedom are not present. By defining a canonically normalized variable

$$F_{in} = \sqrt{\frac{-\Delta m^2}{2\left(-\Delta + m^2\right)}} \tag{3.15}$$

we can bring this action to the form that coincides with the transverse modes of Proca theory:

$$S_V = \frac{1}{2}\int d^4x \left(\dot{F}_{ni}\dot{F}_{ni} - F_{ni,j}F_{ni,j} - m^2F_{ni}F_{ni}\right). \tag{3.16}$$

However, these vector modes are physically more similar to the longitudinal modes of Proca theory. This resemblance can be already noticed in the action for the original fields – the form of the non-trivial kinetic term is the same as in the case of the Proca's longitudinal modes up to a factor of $\frac{1}{2}$. If we would think of considering in addition interactions of $h_{\mu\nu}$, the canonical normalization would introduce singular behavior in mass, similar to the toy model.

3.1.1.3 The Tensor Modes

The action for the tensor modes is given by the following expression:

$$S_T = \frac{1}{8} \int d^4x \left(\dot{h}^T_{ij} \dot{h}^T_{ij} - h^T_{ij,k} h^T_{ij,k} - m^2 h^T_{ij} h^T_{ij} \right) \tag{3.17}$$

Among all of the cases, we can see that this is the simplest one. It describes two tensor degrees of massive graviton.

3.1.2 The VDVZ Discontinuity

Previously, we have seen that the massive gravity has five degrees of freedom – a scalar one, two vector modes, and two tensor modes. The vDVZ discontinuity – the discrepancy between the predictions of massive linearised gravity and General Relativity is due to the longitudinal mode. We will now demonstrate this, following the analysis done in [121].

We will consider a static massive external source, such as Earth, with mass M_0 that is characterized by the 00 component of the energy-momentum tensor $T_{\mu\nu}$. In the static case, the equations for the scalar potentials in the presence of the source are given by:

$$\left(2\Delta - 3m^2\right) \psi = m^2 \Delta E + 2T_{00}, \tag{3.18}$$

$$m^2 \Delta B = 0 \tag{3.19}$$

$$m^2 E = \psi - \phi \tag{3.20}$$

$$\Delta(\phi - \psi) = m^2(\phi - 2\psi - \Delta E) \tag{3.21}$$

From (3.19), it follows that

$$B = 0. \tag{3.22}$$

Substituting (3.20) into (3.21), we obtain

$$\phi = 2\psi. \tag{3.23}$$

Then, (3.18) becomes

$$3\left(\Delta - m^2\right)\psi = 2T_{00}. \tag{3.24}$$

Let us compare these relations to the equations of linearised gravity:

$$\Delta\psi = T_{00} \tag{3.25}$$

$$\psi - \phi = 0 \tag{3.26}$$

$$\Delta(\psi - \phi) = 0. \tag{3.27}$$

We can fix B and E by choosing a gauge $B = E = 0$. Thus, we can right away see that massive and massless theories will give disagreeing predictions – the equations (3.23) and (3.26) differently relate to the two scalar potentials. On the one hand, in the case of linearised gravity, the scalar potential is given by

$$\phi_N = -\frac{M_0}{r}, \tag{3.28}$$

where r is the radial distance from the center of the source. In massive gravity, on the other hand, for $r \ll \frac{1}{m}$, the potential is given by

$$\phi_{mGR} \sim \frac{4}{3}\phi_N. \tag{3.29}$$

Therefore, the perihelion precession of Mercury will not agree in the case of a very small mass and the $m = 0$ case – the gravitational potential is increased in massive gravity by a factor of $\frac{4}{3}$. The bending of light, on the other hand, agrees between the two theories, as it is determined by the combination $\psi + \phi$. However, if we would modify the gravitational constant such that the two gravitational potentials would agree, the theories would nevertheless again differ, now in the results for bending of starlight.

This discontinuity can also be inferred from the propagators. Then, it can be easily seen that the reason why the scalar mode does not decouple lies in the fact that it is coupled to the trace of the energy-momentum tensor [124].

3.1.3 The Non-Linear Terms

Based on the previous results, the massive gravity was initially excluded [59, 61, 62] – the only viable theory seemed to be the one with $m = 0$. Yet, only two years after, this pathology was resolved by A. Vainshtein [63]. He showed that the theory of massive gravity contains a special scale – the Vainshtein radius – given by

$$R_V = \left(\frac{M_0}{m^4}\right)^{\frac{1}{5}}. \tag{3.30}$$

The reason for its appearance are the non-linear terms. The theory that we have considered in 3.1.2 was linearised. However, massive gravity is a non-linear theory. The Vainshtein radius is the scale at which the non-linear terms become of the same order as linear ones. A. Vainshtein has conjectured that once this scale is reached, the longitudinal mode enters a strong coupling regime and decouples from the remaining degrees of freedom for all scales $r < R_v$. Thus, the longitudinal mode would only be effective for $r > R_V$. Moreover, R_V is inversely proportional to the mass of the graviton – assuming that his suggestion is correct, in the massless limit, this scale would rise to infinity, and the predictions of General Relativity would be recovered.

The Dvali-Gabadadze-Porrati (DGP) model – theory in which the matter fields are confined to the four-dimensional brane that is embedded in five-dimensional Minkowski space – was the first model where the Vainshtein conjecture was proved [64, 125]. This theory shares many features with massive gravity – at the lowest order in perturbation theory, the theory has a vDVZ discontinuity. However, in [64] it was shown that the discontinuity is absent in the exact solution of this model, with corrections obtained in [65, 126].

3.1.4 A Connection with Quantum Fluctuations

In order to show that the longitudinal mode enters a strong coupling regime, most of the theories of massive gravity have used an external source (see [124, 127] for a review). Yet, in [67] where a theory of mimetic massive gravity was explored, it was demonstrated that the *essence* of the Vainshtein scale lies in the minimal level of quantum fluctuations.

The state of minimal energy of a classical scalar field corresponds to its zero value. However, when quantized, this is no longer the case – in its ground state, it still fluctuates [42]. The minimal level of quantum fluctuations of a scalar field are a direct consequence of the Heisenberg uncertainty relation. Let us show this by using an estimate as in [38]. We can write the action of a scalar field in a box of volume

L^3 as

$$S \sim \frac{1}{2} \int dt \dot{X}^2, \quad (3.31)$$

where $X = L^{\frac{3}{2}} \delta\varphi_L$. The conjugated momenta P is given by

$$P = \dot{X} \sim \frac{X}{L}. \quad (3.32)$$

Since P and X satisfy Heisenberg uncertainty relation $\Delta P \Delta X \sim 1$, it follows that $\delta\varphi_L \sim \frac{1}{L}$.[1]

Using the minimal amplitude of quantum fluctuations, the authors of [67] were able to deduce which among the self-interacting terms are the most important ones and showed that not only the longitudinal modes become strongly coupled, but that also the vector modes enter a strong coupling regime at a lower scale that followed the strong coupling of the longitudinal modes.

In the following chapters, we will apply this method to the vector theories that we have previously considered. First, we will show that the longitudinal modes of interacting Proca theory, enter a strong coupling regime and demonstrate that beyond the Vainshtein scale one recovers massless theory up to small corrections. Then, in the subsequent chapter, we will analyze the massive Yang-Mills theory and resolve its puzzle of massless limit.

3.2 The Massless Limit of the Interacting Proca Theory

Let us now apply the method of [67] to the interacting Proca theory. By varying the action (2.69) with respect to the vector field, we find the following equations of motion:

$$F^{\nu\mu}_{,\nu} + m^2 A^\mu = \frac{g^2}{2} \left(A_\nu A^{\nu,\mu} - A^\mu A_\nu^{;\nu} \right) \quad (3.33)$$

The case $\mu = 0$ corresponds to the constraint for A_0 (2.94), with the solution given by (2.96). For $\mu = i$, we obtain the equations of motion for the spatial component:

$$\left(\Box + m^2 \right) A_i + A_{j,ij} = \dot{A}_{0,i} + \frac{g^2}{4} \left[\partial_i \left(A_0^2 - A_j A_j \right) - 2 A_i \left(\dot{A}_0 - A_{j,j} \right) \right] \quad (3.34)$$

[1] More rigorously, the minimal amplitude of quantum fluctuations can be computed through an equal time two-point correlation function, as we will see in the next chapter.

Let us now separate the spatial part into the transverse and longitudinal modes according to (2.17). Acting, in addition, with ∂_i on (3.34), we obtain the equation of motion for the longitudinal modes:

$$\begin{aligned}\left(\Box + m^2\right)\chi \sim \frac{g^2}{2m^2}\big[&\ddot{\chi}\Delta\chi - 2\dot{\chi}_{,i}\dot{\chi}_{,i} + \chi_{,ij}\chi_{,ij} + (\Delta\chi)\Box\chi \\ &-\chi_{,i}\Box A_i^T - 2\dot{A}_i^T\dot{\chi}_{,i} + 2A_{i,j}^T\chi_{,ij}\Big],\end{aligned} \tag{3.35}$$

where we have taken the leading order terms for energy scale $k^2 \sim \frac{1}{L^2} \gg m^2$. The next leading order terms to this equation are of the form

$$O\left(\frac{g^2\left(A_i^T\right)^2}{(mL)^2}\right) \quad \text{and} \quad O\left(\frac{g^2\chi^2}{L^2}\right), \tag{3.36}$$

where $\frac{1}{L}$ stands for the derivative. However, these terms are irrelevant on high energies compared to the leading order ones we will quickly see.

In order to find the equation of motion for the transverse modes, we act with the transverse projector

$$P_{ij}^T = \left(\delta_{ij} - \frac{\partial_i\partial_j}{\Delta}\right) \tag{3.37}$$

on (3.34), and find that for energy scales $k^2 \gg m^2$ the equation for the transverse modes with leading order terms is given by

$$\begin{aligned}\left(\Box + m^2\right)A_j^T \sim &-\frac{g^2}{2}P_{ij}^T\left[\left(A_i^T + \chi_{,i}\right)\left(\Box + m^2\right)\chi\right] \\ &-\frac{g^4}{4}P_{ij}^T\left[\chi_{,i}\frac{1}{\Delta}\left(\dot{\chi}_{,k}\dot{\chi}_{,k} + \chi_{,k}\ddot{\chi}_{,k} - \dot{\chi}\Delta\dot{\chi} - \ddot{\chi}\Delta\chi\right)\right]\end{aligned} \tag{3.38}$$

3.2.1 *The Strong Coupling Scale*

3.2.1.1 The Perturbation theory

Let us now develop the perturbation theory for the equations (3.35) and (3.38). For that, we will expand the fields in the powers of g^2:

$$\begin{aligned}\chi &= \chi^{(0)} + \chi^{(1)} + \ldots, \qquad \text{and} \\ A_i^T &= A_i^{T(0)} + A_i^{T(1)} + A_i^{T(2)}\ldots,\end{aligned} \tag{3.39}$$

where, $\chi^{(0)}$ and $A_i^{T(0)}$ satisfy the linear equations of motion

$$\left(\Box+m^2\right)\chi^{(0)}=0 \qquad \text{and} \qquad \left(\Box+m^2\right)A_i^{T(0)}=0, \tag{3.40}$$

whose solutions are plane waves. Then, the first order corrections for the longitudinal modes satisfy the following equation:

$$\begin{aligned}\left(\Box+m^2\right)\chi^{(1)} \sim \frac{g^2}{2m^2}\Big[&-2\dot{\chi}^{(0)}_{,i}\dot{\chi}^{(0)}_{,i}+\chi^{(0)}_{,ij}\chi^{(0)}_{,ij}+\Delta\chi^{(0)}\Delta\chi^{(0)}\\ &-2\dot{A}_i^{T(0)}\dot{\chi}^{(0)}_{,i}+2A^{T(0)}_{i,j}\chi^{(0)}_{,ij}\Big].\end{aligned} \tag{3.41}$$

The first order corrections for the transverse modes satisfy again the free equation:

$$\left(\Box+m^2\right)A_j^{T(1)}=0, \tag{3.42}$$

and the first non-trivial corrections appear at the second order. Taking into account only the most dominant contribution, we find that they satisfy:

$$\left(\Box+m^2\right)A_j^{T(2)} \sim \frac{g^4}{4m^2}P^T_{ij}\left[\chi^{(0)}_{,i}\left(2\dot{\chi}^{(0)}_{,k}\dot{\chi}^{(0)}_{,k}-\Delta\chi^{(0)}\Delta\chi^{(0)}-\chi^{(0)}_{,kl}\chi^{(0)}_{,kl}\right)\right] \tag{3.43}$$

The strong coupling scale – the scale that emerges once the non-linear terms become of the order of linear ones – signalizes the breakdown of the perturbation theory. Let us look to which scale it corresponds in the case of the longitudinal modes. As a first step, we have to determine what are the most dominant terms in (3.41). The minimal level of quantum fluctuations allows us to do precisely that.

3.2.1.2 The Minimal Level of Quantum Fluctuations

In the case of the normalised longitudinal and transverse modes, the size of the quantum fluctuations can be evaluated using the equal-time two point correlation functions [42]:

$$\zeta_{\chi_n}(|\boldsymbol{x}-\boldsymbol{y}|) \equiv \langle 0|\,\hat{\chi}_n(\boldsymbol{x},t)\hat{\chi}_n(\boldsymbol{y},t)\,|0\rangle \tag{3.44}$$

$$\zeta_{A^T}(|\boldsymbol{x}-\boldsymbol{y}|) \equiv \langle 0|\,\hat{A}_i^T(\boldsymbol{x},t)\hat{A}^{Ti}(\boldsymbol{y},t)\,|0\rangle \tag{3.45}$$

Inserting the operator expansion (2.36) we obtain the following expression

$$\zeta_{\chi_n}(r)=\int\frac{dk}{(2\pi)^2 k}\frac{k^3}{\omega_{\mathbf{k}}}\frac{\sin(kr)}{kr} \tag{3.46}$$

for the longitudinal modes, and

$$\zeta_{A^T}(r) = \int \frac{dk}{(2\pi)^2 k} \frac{k^3}{\omega_{\mathbf{k}}} \frac{\sin(kr)}{kr} \tag{3.47}$$

for the transverse modes. Here, $r = |\boldsymbol{x} - \boldsymbol{y}|$. For the scales $L \sim \frac{1}{k}$, $\frac{sin(kr)}{kr} \sim O(1)$. Thus, the minimal level of quantum fluctuations in the scales $\frac{1}{L} \gg m$ for the transverse and normalised longitudinal modes respectively are given by:

$$\delta A_L^T \sim \frac{1}{L} \qquad \delta \chi_{nL} \sim \frac{1}{L} \tag{3.48}$$

Taking into account that the normalised longitudinal mode is given in terms of the original mode by (2.33), we obtain the following result for the longitudinal mode:

$$\delta \chi_L \sim \frac{1}{Lm}. \tag{3.49}$$

3.2.1.3 The Breakdown of the Perturbation Theory

Let us now use the minimal level of quantum fluctuations to find the strong coupling scale. Estimating the derivatives as $\partial_\mu \sim \frac{1}{L}$, we can see that the following non-linear terms in the equation (3.41) are of the same order

$$\frac{g^2}{2m^2} \dot{\chi}_{,i}^{(0)} \dot{\chi}_{,i}^{(0)} \sim \frac{g^2}{2m^2} \chi_{,ij}^{(0)} \chi_{,ij}^{(0)} \sim \frac{g^2}{2m^2} \Delta\chi^{(0)} \Delta\chi^{(0)} \sim \frac{g^2}{m^2} \frac{\left(\chi^{(0)}\right)^2}{L^4}, \tag{3.50}$$

and

$$\frac{g^2}{2m^2} \dot{A}_i^{T(0)} \dot{\chi}_{,i}^{(0)} \sim \frac{g^2}{2m^2} A_{i,j}^{T(0)} \chi_{,ij}^{(0)} \sim \frac{g^2}{m^2} \frac{A^{T(0)} \chi^{(0)}}{L^3}. \tag{3.51}$$

Taking into account the minimal amplitude of quantum fluctuations for the transverse and the original longitudinal modes, we can further evaluate these terms as

$$\frac{g^2}{m^2} \frac{\left(\chi^{(0)}\right)^2}{L^4} \sim \frac{g^2}{(mL)^4 L^2} \quad \text{and} \quad \frac{g^2}{m^2} \frac{A^{T(0)} \chi^{(0)}}{L^3} \sim \frac{g^2}{(mL)^3 L^2} \tag{3.52}$$

Since we are considering length-scales $\frac{1}{L^2} \gg m^2$, clearly, the first term is the leading one. Therefore, we can evaluate the leading corrections to the longitudinal modes as

$$\chi^{(1)} \sim \frac{g^2}{(mL)^4}. \tag{3.53}$$

The perturbation theory for the longitudinal modes breaks down when this term becomes of the same order as the linear term, $\chi^{(0)}$. Evaluating it as $\chi^{(0)} \sim \frac{1}{mL}$, we find that the strong coupling scale for the longitudinal modes is given by

$$L_{str} \sim \frac{g^{2/3}}{m}. \tag{3.54}$$

Therefore, for the length scales $L < L_{str}$, the perturbation theory for the longitudinal modes is no longer trustable.

Nonetheless, we should still check what happens with the transverse modes. The leading corrections can be evaluated as

$$A_j^{T(2)} \sim \frac{g^4}{(mL)^5 L}. \tag{3.55}$$

The scale at which these corrections become of the same order as the linear term, $A_j^{(0)} \sim \frac{1}{L}$ is given by

$$L^T \sim \frac{g^{4/5}}{m}. \tag{3.56}$$

At the first sight, this suggests that the transverse modes enter the strong coupling regime – a surprising, nonphysical result, as it would mean that if the photon has a small mass and can self-interact, our phones could no longer work. However, we can notice that this scale is smaller than the one for the longitudinal modes – the breakdown of the perturbation theory for them appears before we can reach the scale L^T. Therefore, this result is not trustworthy. In order to find out what is actually happening to the transverse modes, we must go beyond the strong coupling scale.

However, before we do it, we should check that the assumption (2.95) holds within the perturbation theory – if not, we would need to find another way to express the theory in terms of only the longitudinal and transverse modes in the first place. We can evaluate the assumption as

$$g^2 \chi \ll 1 \tag{3.57}$$

Then, estimating $\chi \sim \frac{1}{mL}$, it becomes

$$g^{4/3} \frac{L_{str}}{L} \ll 1. \tag{3.58}$$

The perturbation theory for the longitudinal modes is valid for scales $L > L_{str}$. Therefore, since the coupling constant is smaller than unity, the assumption is satisfied.

3.2.2 *Beyond the Strong Coupling Scale*

We have seen that at scales $L \sim L_{str}$ the longitudinal mode becomes strongly coupled. For $L < L_{str}$, the kinetic terms are suppressed by the following non-linear terms

$$\mathcal{L}_{1\chi} \sim \frac{g^2}{2}\left(\dot{\chi}^2 - \frac{1}{2}\chi_{,i}\chi_{,i}\right)(-\Delta\chi) \tag{3.59}$$

From these, we can evaluate the new minimal level of quantum fluctuations for the longitudinal modes which will allow us to go beyond the strong coupling scale. First, we can notice that the canonical normalisation which was determined by the kinetic term for scales $k < k_{str}$ now changes. Substituting,

$$d\tilde{\chi}_n = g\sqrt{-\Delta\chi}\,d\chi \tag{3.60}$$

the previous Lagrangian density reduces to

$$\mathcal{L}_{1\chi} \sim \frac{1}{2}\left(\dot{\tilde{\chi}}_n^2 - \frac{1}{2}\tilde{\chi}_{n,i}\tilde{\chi}_{n,i}\right) \tag{3.61}$$

This means that the minimal level of quantum fluctuations for the normalised longitudinal mode on the scales $k^2 \sim \frac{1}{L^2} \gg m^2$ is given by $\delta\tilde{\chi}_{nL} \sim \frac{1}{L}$. Furthermore, this implies that the minimal level of quantum fluctuations of the original mode is

$$\delta\chi_L \sim \frac{1}{g^{\frac{2}{3}}} \tag{3.62}$$

which agrees with the quantum fluctuations of the longitudinal mode on the strong coupling scale, i.e. $\delta\chi_{L_{str}} \sim \frac{1}{mL_{str}} \sim \frac{1}{g^{\frac{2}{3}}}$.

It is now easy to see that the non-linear terms for the longitudinal modes will dominate for all $k > k_{str}$. For example, the most dominating terms and the interacting terms considered before can be evaluated as

$$\frac{g^2}{2}\dot{\chi}^2(-\Delta\chi) \sim \frac{1}{L^4} \qquad -\frac{g^2}{2}A_i^T\chi_{,i}\Box\chi \sim \frac{g^{\frac{2}{3}}}{L^4} \qquad -\frac{g^2}{4}A_i^{T^2}\Box\chi \sim \frac{g^{\frac{4}{3}}}{L^4} \tag{3.63}$$

Since $g \ll 1$ the first term is clearly dominating. The next non-linear term would come as

$$\frac{g^4\chi^4}{L^4} \sim \frac{g^{\frac{4}{3}}}{L^4} \tag{3.64}$$

However, we see that this one is also always smaller than the leading one. Similar can be shown for all higher-order terms. Furthermore, the initial assumption still holds

$$g^2\chi \sim g^{\frac{4}{3}} \ll 1. \tag{3.65}$$

Let us now find out what happens to the transverse modes. The perturbative approach had suggested that they also enter a strong coupling regime, albeit on smaller scales than the longitudinal modes. Therefore, at the strong coupling scale and length scales below, it will still be possible to use the perturbation theory for the transverse modes. However, in contrast to the previous case, the leading order corrections to the transverse modes that arise due to the longitudinal ones are now given by

$$\left(\Box + m^2\right) A_j^{T(1)} \sim -\frac{g^2}{2} P_{ij}^T \left[\chi_{,i} \left(\Box + m^2\right) \chi\right], \tag{3.66}$$

and can be evaluated as

$$A_j^{T(1)} \sim g^2 \frac{\chi^2}{L^3} \sim \frac{g^{2/3}}{L^3}. \tag{3.67}$$

Therefore, the transverse modes do not enter the strong coupling regime. Even though the perturbation theory for the longitudinal modes breaks down at L_{str}, it is still valid for the transverse modes.

Moreover, when the mass goes to zero, the strong coupling scale rises to infinity and we fully recover the corresponding massless theory – the massless limit of this toy model is smooth.

We can notice that the corrections due to the strongly coupled mode remain outside the perturbation theory. Even though this might come as a surprise, it is due to the nature of the self-interaction we have considered. The self-interaction is not gauge invariant and thus, the corrections would remain even in the massless theory. Nevertheless, once the longitudinal mode enters the strong coupling regime, it loses its linear propagator and ceases to evolve, similarly to the Vainshtein mechanism in massive gravity where the longitudinal mode does not propagate beyond the Vainshtein radius.

3.2.3 *The Higher-Order Terms*

For completeness, let us show that the higher order terms are always subdominant when compared to the leading order terms for the transverse and longitudinal modes.

We can represent them as

$$\mathcal{L}_{int} \sim \sum_{n=0}^{\infty} g^{4+2n} \left(\frac{\chi^{4+n}}{L^4} + \frac{\chi^{3+n}}{L^3} A^T + \frac{\chi^{2+n}}{L^2} \left(A^T\right)^2 + \frac{\chi^{1+n}}{L} \left(A^T\right)^3 + \chi^n \left(A^T\right)^4 \right) \tag{3.68}$$

For the scales $L > L_{str}$, the most dominant term for the transverse modes corresponds to

$$g^{4+2n} \frac{\chi^{3+n}}{L^3} A^T. \tag{3.69}$$

Therefore, let us show that this term is the most dominant one among all possible higher order terms. For $L > L_{str}$, we can evaluate

$$\chi \sim \frac{1}{mL} \quad \text{and} \quad A^T \sim \frac{1}{L}. \tag{3.70}$$

For each $n = 0, 1, 2, ...$ it is clear that the terms with most longitudinal modes are dominating:

$$g^{4+2n} \frac{\chi^{4+n}}{L^4} \tag{3.71}$$

for the contribution to the longitudinal modes, and

$$g^{4+2n} \frac{\chi^{3+n}}{L^3} A^T \tag{3.72}$$

for the contribution to the transverse modes. This is also the case for $L < L_{str}$, where we can evaluate the longitudinal and transverse modes as

$$\chi \sim \frac{1}{g^{2/3}} \quad \text{and} \quad A^T \sim \frac{1}{L}. \tag{3.73}$$

Now all we have to make sure of is that with increasing n, the remaining terms give a smaller contribution. This is the case if the following holds:

$$g^{4+2n} \frac{\chi^{3+n}}{L^3} A^T > g^{4+2(n+1)} \frac{\chi^{4+(n+1)}}{L^4} \tag{3.74}$$

or equivalently,

$$1 > g^2 \chi^2 \tag{3.75}$$

For $L > L_{str}$, this condition is equivalent to

$$1 > g^{2/3}\left(\frac{L_{str}}{L}\right)^2 \tag{3.76}$$

which is clearly satisfied. This is the case also for $L < L_{str}$, where it is equivalent to

$$1 > g^{2/3}. \tag{3.77}$$

Therefore, among the infinite number of terms in (3.68), only (3.71) and (3.72) are relevant. However, using the same arguments, we can show then that (3.71) is subdominant in comparison to $L_{g\chi}$ in equation (2.97), while for $L < L_{str}$, term (3.72) given a smaller contribution than $L_{gA_i^T\chi}$.

Chapter 4
The Massless Limit of Massive Yang-Mills Theory

The standard approaches suggest that the massless limit of massive Yang-Mills theory, with mass added *by hand* is not smooth—the perturbative series are singular in mass. This implies that the unitarity is violated once the energy scales $k_u \sim \frac{m}{m}$ is reached. In the first chapter, we have confirmed that the source of this discontinuity are the longitudinal modes.

In [58], it was conjectured that beyond the perturbation theory, the massless limit is smooth. Yet, to date, a proof of this conjecture was nevertheless missing—the corrections beyond the perturbative approach were unknown. Moreover, subsequent studies have shown that a non-polynomial Lagrangian obtained in [58] can be algebraically reduced to the polynomial form through algebraic re-definitions of fields that leave the S-matrix unchanged. The conclusion opposed to that of [58]—the resulting theory seemed to be discontinuous in the massless limit. Yet, the two approaches have differed in the definition of the degrees of freedom—the first one defined them non-linearly, while the second was equivalent to a linear decomposition. The two possibilities, however, do not change the S-matrix, thus complicating the puzzle further as it was unclear which decomposition to choose. Hence, we are led to the question—*Is there an underlying mechanism that might resolve these puzzles?*

In the previous chapter, we have seen that the Vainshtein mechanism resolves the discontinuous predictions of massive linearised gravity. The non-linear terms have caused the longitudinal mode to enter a strong coupling regime at the Vainshtein radius. Beyond it, it decouples from the remaining degrees of freedom, restoring the agreement of massive gravity and General Relativity. Moreover, we have applied this mechanism also to the toy model, and have seen that a similar situation arises. The purpose of this chapter is to show that the Vainshtein mechanism answers the puzzles of the massive Yang-Mills theory. We will not only see that the only correct definition of the degrees of freedom is a non-linear one—a conclusion that is not possible to reach with the standard methods—but we will also find that the longitudinal modes enter a strong coupling regime at the scale that matches with that

A. Hell, *The Massless Limit of Massive Gauge Theories*, Springer Theses,
https://doi.org/10.1007/978-3-032-01090-2_4

of the unitarity violation. Beyond it, they will decouple from the remaining degrees of freedom—the transverse modes—restoring the massless theory up to very small corrections that will become even smaller as we approach higher energies.

4.1 The Linear Decomposition and the Strong Coupling Scales

We will start the analysis of the massive Yang-Mills theory by rewriting it in terms of the propagating degrees of freedom—the transverse and longitudinal modes. Initially, we will linearly decompose the spatial part of the vector field:

$$A_i^a = A_i^{Ta} + \chi_{,i}^a, \tag{4.1}$$

as in Sect. 2.4.2. In this case, we have previously found at the following Lagrangian density

$$\mathcal{L} = \mathcal{L}_0 + \mathcal{L}_{int}, \qquad \text{where} \tag{4.2}$$

$$\mathcal{L}_0 = -\frac{1}{2}\chi^a(\Box + m^2)\frac{m^2(-\Delta)}{-\Delta + m^2}\chi^a - \frac{1}{2}A_i^{Ta}(\Box + m^2)A_i^{Ta} \qquad \text{and}$$

$$\mathcal{L}_{int} \sim g\varepsilon^{abc}\left[\frac{1}{2}\chi^b\chi_{,i}^c\Box A_i^{Ta} - \dot{\chi}^a\chi_{,i}^c\frac{m^2}{\Delta}\left(\dot{\chi}_{,i}^b\right) + \chi^b A_i^{Tc}\Box A_i^{Ta}\right]$$
$$+\frac{g^2}{2}\varepsilon^{fab}\varepsilon^{fcd}\left[\dot{\chi}^a\dot{\chi}^c\chi_{,i}^b\chi_{,i}^d - \frac{1}{2}\chi_{,i}^a\chi_{,i}^c\chi_{,j}^b\chi_{,j}^d\right.$$
$$\left.+\left(\dot{\chi}_{,i}^a\chi_{,i}^b + \dot{\chi}^a\Delta\chi^b\right)\frac{1}{\Delta}\left(\dot{\chi}_{,j}^c\chi_{,j}^d + \dot{\chi}^c\Delta\chi^d\right)\right],$$

after resolving the constraint satisfied by the temporal component of the vector field. Following the previous chapter, we will now explore the behaviour of the transverse and longitudinal modes and look for the scale at which perturbative approach breaks down. We have learned that the minimal amplitudes of quantum fluctuations play an important role there—they will be neccessary to compare different terms and determine the Vainshtein scale. From the kinetic terms, we can infer that for the original fields, they are given by:

$$\delta A_L^{Ta} \sim \frac{1}{L} \qquad \text{and} \qquad \delta\chi_L^a \sim \frac{1}{mL}, \qquad a = 1, 2, 3. \tag{4.3}$$

Let us remember that with the canonically normalized longitudinal modes, we could already infer from (2.114), that due to them, inverse powers of mass appear in the interactions indicating a discontinuity of the massless limit. Hence, we might suspect by taking the analogy with massive gravity that the longitudinal modes will enter the strong coupling regime. Nevertheless, in the case of a linear decomposition (4.1), this question is much more subtle.

The equations of motion for the longitudinal modes are given by

$$\begin{aligned}\left(\Box + m^2\right)\chi^a \sim \frac{g}{m^2}\varepsilon^{abc}\Big[&\chi^b_{,i}\Box A^{Tc}_i + A^{Tb}_i\Box A^{Tc}_i \\ &+m^2\chi^c_{,i}\frac{1}{\Delta}\ddot{\chi}^b_{,i} + \frac{m^2}{\Delta}\left(\ddot{\chi}^b_{,i}\chi^c_{,i} + \ddot{\chi}^b\Delta\chi^c + \dot{\chi}^b\Delta\dot{\chi}^c\right)\Big] \\ &+\frac{g^2}{m^2}\varepsilon^{fab}\varepsilon^{fcd}\chi^b_{,i}\left[\frac{\partial_i\partial_0}{\Delta}\left(\dot{\chi}^c_{,j}\chi^d_{,j} + \dot{\chi}^c\Delta\chi^d\right) - \dot{\chi}^c\dot{\chi}^d_{,i}\right. \\ &\left.-\ddot{\chi}^c\chi^d_{,i} + \chi^d_{,i}\Delta\chi^c + \chi^c_{,j}\chi^d_{,ij}\right],\end{aligned} \tag{4.4}$$

while for the transverse modes, we obtain

$$\left(\Box + m^2\right)A^{Ta}_i \sim g\varepsilon^{abc}P^T_{ij}\left[\frac{1}{2}\Box\left(\chi^b\chi^c_{,j}\right) + A^{Tc}_j\Box\chi^b + 2\chi^b_{,\mu}A^{Tc,\mu}_j\right]. \tag{4.5}$$

Both of these equations contain only the most important terms that we will now study. Let us develop perturbation theory, by expanding the longitudinal and transverse modes in powers of the coupling constant:

$$A^T_i = A^{T(0)}_i + A^{T(1)}_i + \ldots \qquad \text{and} \qquad \chi = \chi^{(0)} + \chi^{(1)} + \ldots, \tag{4.6}$$

where $A^{T(0)}_i$ and $\chi^{(0)}$ satisfy linear equations

$$\left(\Box + m^2\right)A^{Ta(0)}_i = 0 \qquad \text{and} \qquad \left(\Box + m^2\right)\chi^{a(0)} = 0, \tag{4.7}$$

whose solutions are plane waves. In the case of the longitudinal modes, we will be interested in corrections arising at both $\mathcal{O}(g)$ and $\mathcal{O}(g^2)$. Therefore, we will first analyze the leading order corrections to the transverse modes, as these will play the main role in the higher-order corrections.

4.1.1 The Strong Coupling of the Transverse Modes

The leading order corrections to the transverse modes are determined by the following relation:

$$\left(\Box + m^2\right)A^{Ta(1)}_i \sim g\varepsilon^{abc}P^T_{ij}\left(\chi^{b(0)}_{,\mu}\chi^{c(0),\mu}_{,j} + 2\chi^{b(0),\mu}A^{Tc(0)}_{i,\mu}\right). \tag{4.8}$$

Estimating $\partial_\mu \sim \frac{1}{L}$, we can evaluate the non-linear terms as

$$g\varepsilon^{abc} P^T_{ij} \left(\chi^{b(0)}_{,\mu} \chi^{c(0),\mu}_{,j}\right) \sim \frac{g\chi^2}{L^3} \qquad \text{and}$$
$$g\varepsilon^{abc} P^T_{ij} \left(\chi^{b(0),\mu} A^{Tc(0)}_{i,\mu}\right) \sim \frac{g\chi A^T}{L^2}. \tag{4.9}$$

Moreover, using (4.3), we can further evaluate them as

$$\frac{g\chi^2}{L^3} \sim \frac{g}{(mL)^2 L^3}, \tag{4.10}$$

and

$$\frac{g\chi A^T}{L^2} \sim \frac{g}{(mL) L^3}. \tag{4.11}$$

From here, it follows that the first term is most dominant one—we are evaluating the theory for length scales $\frac{1}{L^2} \gg m^2$ as for these scales the theory is problematic in the massless limit. Thus, the leading order corrections can be evaluated as

$$A^{Ta(1)}_i \sim \frac{g}{(mL)^2 L}. \tag{4.12}$$

The strong coupling scale is the scale at which these non-linear terms, become of the same order as the linear ones, i.e.

$$A^{Ta(1)}_i \sim A^{Ta(0)}_i. \tag{4.13}$$

Evaluating $A^{Ta(0)}_i \sim \frac{1}{L}$, we therefore find that the transverse modes enter a strong coupling regime at lenght-scales

$$L^T_{str} \sim \frac{\sqrt{g}}{m}. \tag{4.14}$$

Yet, this result is surprising—the length scale L^T_{str} is larger than the one corresponding to the breakdown of the unitarity, that is given by

$$L_u \sim \frac{g}{m}. \tag{4.15}$$

This means that the perturbation theory in transverse modes breaks down before unitarity is violated. However, the fact that the transverse modes enter a strong coupling regime is an unphysical result. Loosely speaking, it would mean that if we would add even a small mass to the gluons, we could forget about the weak interactions. The only possibility that could aid this scenario is if the longitudinal modes enter a strong coupling regime on lower or equal lenght scales. Therefore, before we draw any other conclusions, let us study the corrections to the longitudinal modes.

4.1.2 *The Longitudinal Modes*

Using the equation of motion for the longitudinal modes (4.4), we can find that the first-order corrections to the longitudinal modes are given by

$$\left(\Box + m^2\right) \Delta\chi^{a(1)} \sim g\varepsilon^{abc}\chi^{b(0)}_{,\mu}\Delta\chi^{c(0),\mu}. \tag{4.16}$$

We can estimate these corrections as

$$\chi^{a(1)} \sim \frac{g}{(mL)^2}. \tag{4.17}$$

Once they become of the same order as the linear term, i.e.

$$\chi^{(1)} \sim \chi^{(0)}, \tag{4.18}$$

the perturbation theory in the longitudinal modes breaks down—they become strongly coupled. This corresponds to the following scale

$$L_{str} \sim \frac{g}{m}. \tag{4.19}$$

However, this scale agrees with that of unitarity violation. It is smaller than the previously found scale L^T_{str} at which the transverse modes have entered a strong coupling regime.

At the first sight, (4.4) suggests that the only term that might still give rise to the same strong coupling scale is present at the higher-order. As a next step, we will study these corrections in two cases—with and without the presence of the transverse modes.

4.1.2.1 Case 1: $A^T_i = 0$

Let us first set the transverse modes to zero. Then, the second-order corrections are given by

$$\left(\Box + m^2\right) \chi^{a(2)} \sim -\frac{g^2}{m^2}\varepsilon^{fab}\varepsilon^{fcd} P^T_{ij}\left(\chi^{d(0),\mu}_{,j}\chi^{c(0)}_{,\mu}\right), \tag{4.20}$$

and we can estimate them as

$$\chi^{(2)} \sim \frac{g^2}{(mL)^5}. \tag{4.21}$$

The scale at which these non-linear terms become of the same order as the linear term is given by

$$L^T_{str} \sim \frac{\sqrt{g}}{m}, \tag{4.22}$$

corresponding to the same scale that we have found when analysing the transverse modes. For scales $L < L^T_{str}$, the most relevant terms in the Lagrangian density are given by

$$\begin{aligned} L_{int} \supset \frac{g^2}{2}\varepsilon^{fab}\varepsilon^{fcd}\Big[&\dot{\chi}^a\dot{\chi}^c\chi^b_{,i}\chi^d_{,i} - \frac{1}{2}\chi^a_{,i}\chi^c_{,i}\chi^b_{,j}\chi^d_{,j} \\ &+\left(\dot{\chi}^a_{,i}\chi^b_{,i} + \dot{\chi}^a\Delta\chi^b\right)\frac{1}{\Delta}\left(\dot{\chi}^c_{,j}\chi^d_{,j} + \dot{\chi}^c\Delta\chi^d\right)\Big]. \end{aligned} \tag{4.23}$$

For simplicity, let us evaluate them as We can estimate these terms as

$$\mathcal{L}_{int} \supset \frac{g^2\chi^4}{L^4}. \tag{4.24}$$

Following the approach of the toy model, with their help, we can determine the new minimal amplitude of quantum fluctuations for the original longitudinal modes. Defining the new canonically normalised variable $\chi_n \sim \frac{g}{L}\chi^2$, we find that within the strong coupling regime, they are given by

$$\delta\chi^a_L \sim \frac{1}{\sqrt{g}}. \tag{4.25}$$

Therefore, in the absence of the transverse modes, the longitudinal mode enters the strong coupling regime at length-scales L^T_{str}. Let us now investigate what changes when the transverse modes are included back in the theory.

4.1.2.2 Case 2: $A^T_i \neq 0$

In this case, the equations satisfied by the second-order corrections contain another term, given by:

$$\left(\Box + m^2\right)\chi^{a(2)} \sim \frac{g}{m^2}\varepsilon^{abc}\chi^{b(0)}_{,i}\Box A^{Tc(1)}_i - \frac{g^2}{m^2}\varepsilon^{fab}\varepsilon^{fcd}P^T_{ij}\left(\chi^{d(0),\mu}_{,j}\chi^{c(0)}_{,\mu}\right) \tag{4.26}$$

By substituting (4.8) into (4.26), the second term cancels.

4.1.3 From Linear to a Nonlinear Decomposition

At a first glance, the previous results indicate that the only degrees of freedom that are entering the strong coupling regime are transverse modes. The perturbation theory suggests that the longitudinal modes, on the other hand, become strongly coupled at a lower scale, which agrees with that of the unitarity violation.

However, this conclusion is partially incorrect. As we have found, the transverse modes indeed become strongly coupled. However, this means that the perturbation theory for them breaks down. In other words, the second-order corrections for the longitudinal modes were not correctly evaluated, as, at that point, the relation (4.8) is no longer valid. Thus, the second term in (4.26) cannot be cancelled for scales $L \geq L^T_{str}$.

At the strong coupling scale L^T_{str}, we can therefore evaluate the second order contributions to the longitudinal modes as

$$\chi^{(2)} \sim \frac{g^2}{\left(mL^T_{str}\right)^5}. \tag{4.27}$$

As a result, at the scale L^T_{str} the longitudinal modes also enter a strong coupling regime—not only is the above term more dominant than the first-order correction but is it also of the same order as the linear term.

Below this scale, the most dominant terms in the interacting part of the Lagrangian density are given by

$$\mathcal{L}_{int} \sim \frac{g}{2}\varepsilon^{abc}\chi^b\chi^c_{,i}\Box A^{Ta}_i + \frac{g^2}{2}\frac{\chi^4}{L^4}. \tag{4.28}$$

Let us use these terms to determine the behavior of the transverse and the longitudinal modes. We can see that the last term coincides with 4.1.2.1. With it, we can determine that the minimal level of the quantum fluctuations of the longitudinal modes are given by (4.25).

The behaviour of the transverse modes is determined by the first term. Using (4.25), we can evaluate it as

$$\frac{g}{2}\varepsilon^{abc}\chi^b\chi^c_{,i}\Box A^{Ta}_i \sim \frac{1}{L^3}A^T. \tag{4.29}$$

Thus, in contrast to the longitudinal modes, the minimal level of quantum fluctuations for the transverse modes beyond the strong coupling is left unchanged, and given by

$$\delta A^T_{L \leq L^T_{str}} \sim \frac{1}{L}. \tag{4.30}$$

Thus, the quantum fluctuations of the transverse modes are left unchanged, while those for longitudinal modes become constant after the strong coupling scale is crossed. However, we can also notice that the first term of (4.28) remains of the same order as the kinetic term in the transverse modes. There is only one possibility that can therefore resolve the ambiguity of the strongly coupled transverse modes—for scales $L \leq L^T_{str}$, they are no longer properly defined. This is also apparent from the equations of motion for the transverse modes (4.31). By taking into account only the most important terms, we can rewrite it as:

$$\Box\left[A_i^{Ta} - \frac{g}{2}\varepsilon^{abc}P^T_{ij}\left(\chi^b\chi^c_{,j}\right)\right] \sim g\varepsilon^{abc}P^T_{ij}\left[A_j^{Tc}\Box\chi^b + 2\chi^b_{,\mu}A_j^{Tc,\mu}\right]. \tag{4.31}$$

The two terms on the left-hand side are of the same order. As a result, we should redefine the transverse modes as

$$A_i^{Ta} = B_i^{Ta} + \frac{g}{2}\varepsilon^{abc}P^T_{ij}\left(\chi^b\chi^c_{,j}\right), \qquad B^{Ta}_{i,i} = 0, \tag{4.32}$$

and repeat the previous procedure for the new transverse modes B_i^{Ta}.

Once this replacement has been made, the non-linear terms that have introduced the strong coupling scale L^T_{str} drop out of the Lagrangian. Nonetheless, another strong coupling scale will appear

$$\tilde{L}^T_{str} \sim \frac{g^{2/3}}{m}, \tag{4.33}$$

this time lower than L^T_{str}, but still larger than the scale at which the unitarity breaks down. It will then require another redefinition and the repetition of the whole process.

Therefore, even if we start with a linear definition of the transverse and longitudinal modes, we will nevertheless end up with a non-linear one. It will turn out that one will encounter an infinite ladder of strong coupling scales, all approaching the unitarity violation scale, and that will require an infinite number of field redefinitions. Unfortunately, we do not have an infinite amount of time, however, luckily, in the following we will deduce the correct non-linear decomposition, resulting in a non-polynomial Lagrangian.

This clarifies the conclusions of [119] that have arisen by analyzing a Lagrangian with polynomial terms. Such a Lagrangian corresponds to the linearly defined transverse and longitudinal modes. However, the strong coupling analysis shows that even if we would start with such a decomposition, it would soon fail, and we would end up with a non-linear one.

4.1.4 Diagrammatic Techniques and L^T_{str}

The previous results might still leave us puzzled. In the literature, the unitarity breaks down at length scales $L_u \sim \frac{g}{m}$—this can be inferred from the two-loop corrections to the propagator of the transverse modes [54]. However, in the case of a linear decomposition, we have found that the transverse and longitudinal modes enter a strong coupling regime much before, at scales $L_T \sim \frac{\sqrt{g}}{m}$. This was found on the level of equations of motion for the transverse modes at the lowest order in the coupling constant, that afterward implied the strong coupling of the longitudinal modes also, at the higher-order, as the perturbation theory in transverse modes ceased to be valid. Hence, it is natural to suspect that the same scale could be inferred also by using diagrammatic techniques. In particular, the question that we will focus on is if the unitarity violation scale could be lower than stated in the literature. However, by studying the two-loop corrections to the propagator of the transverse modes, we will show that this does not seem to be the case (Fig. 4.1).

At two loops, the two diagrams that could give us a scale that matches with L^T_{str} are the following ones:

Fig. 4.1 Two-loop corrections to the propagator of the transverse modes in massive Yang-Mills theory

By counting the number of longitudinal propagators, it is easy to see that the maximal singularity in mass that these diagrams could have is

$$\frac{1}{m^8}, \tag{4.34}$$

indicating a possibility for the existence of a unitarity violation scale that matches with L^T_{str}. However, this will turn out not to be the case. To show it, we need to confirm that at least one of the vertices is proportional to the square of the mass in the leading order. Let us now analyze the imaginary part of these diagrams (Figs. 4.2 and 4.3).

The cuts of the first diagram are given by

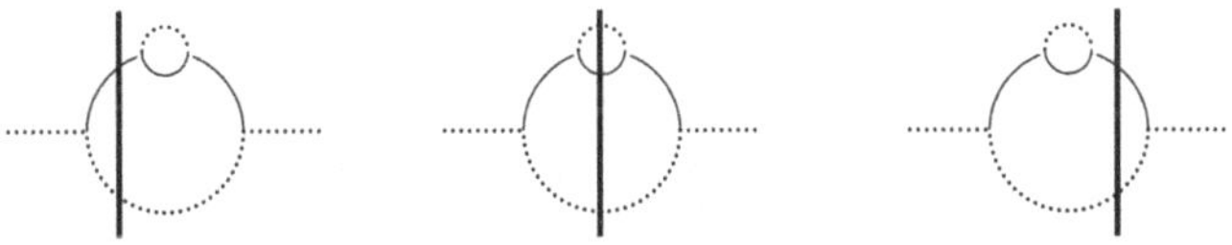

Fig. 4.2 The cuts of the first two-loop diagram

while for the second diagram we have

Fig. 4.3 The cuts of the second two-loop diagram

Now, let us examine the first diagram. We will denote by p the external moments and by k and l the internal moments over which we are integrating, with k corresponding to the lowest line and l to the line above. As a first step, we will insert the following relation

$$\int d^4k \int d^4l = \int d^4k \int d^4l \int d^4q\delta^{(4)}(p-k-l-q) \tag{4.35}$$

into the expression for the imaginary part of the diagram. As cutting a line means that the corresponding momenta are set on-shell, we will have $k^2 = m^2$ in all three cuts. Next, let us look at one of the vertices appearing in the last two diagrams. Using the fact that the external momentum is on-shell, i.e. $\varepsilon_i^a(p)p_i = 0$ and $p^2 = m^2$, the first vertex of the middle diagram is given by the following

$$V^{abc}_{i,2\chi}(-p_{(i)}, k, p-k) \sim g\varepsilon^{abc}\left(-p_\mu k^\mu p_i - p^2 k_i\right) = -g\varepsilon^{abc}m^2 k_i, \tag{4.36}$$

in the leading order. In the case of the last diagram, the equivalent relation holds for the last vertex, instead of the first one. Thus clearly, by analysing the cuts of the first diagram we can see that it is not possible to infer from them a scale that matches L^T_{str}.

Let us now look at the second diagram. To analyze the cuts, we will first perform the same trick (4.35) as before. Due to the first, or equivalently last vertex over which the cut is made, the last two cuts cannot provide the scale L^T_{str}—the cuts set all three lines around the vertex on-shell, and as a result, only the terms proportional to m^2 will appear in the vertex.

For the second diagram, the first two cuts are equivalent again. The cut through the line that passes through the loop leaves us with the transverse projector in the momentum space. However, this expression is zero:

$$l_i\left(\delta_{ij} - \frac{l_i l_j}{|\boldsymbol{l}|^2}\right) = 0, \tag{4.37}$$

where l corresponds to this line. Therefore, now all that we have to consider is the vertex at the bottom of this line. This vertex is given by

$$V^{abc}_{i,2\chi}(l_{(i)}, k, -l-k) \sim g\varepsilon^{abc} l_\mu \left(l^\mu k_i - k^\mu l_i\right), \tag{4.38}$$

where k is the loop momentum that exits the vertex. Due to the contraction with the transverse projector, the last term vanishes. Moreover, as the cut is made through the middle line, it sets the momentum l on-shell. From here, it follows that this vertex also has a term proportional to the mass. Thus, we can see that the analysis of the two-loop corrections for the transverse modes suggests that L^T_{str} does not appear in the diagrams—its origin seems to lie only in the strong coupling.

4.2 A Key to the Non-linear Decomposition

So far we have seen that if we start with linearly defined transverse and longitudinal modes, we will necessarily end up with a non-linear definition. This is due to the behavior of the transverse modes beyond the first strong coupling scale. There, the non-linear and linear terms of the transverse modes remain of the same order, indicating a new definition of these degrees of freedom. After such redefinition, we would nevertheless end up with a similar situation—a new strong coupling scale would appear, requiring another redefinition. Ultimately, an infinite number of re-definitions would be necessary, until the strong coupling scale would match with the scale at which unitarity is violated. However, the previous procedure is not ideal to find the final decomposition. Thus, we must search for a different approach that would allow us to find it. A key that will help us to achieve this goal lies in the gauge-invariant formulation of the massless Yang-Mills theory.

4.2.1 A Trouble with the Linear Decomposition

Yang-Mills theory is a theory of a massless non-Abelian vector field that has only two degrees of freedom per color—the transverse modes [98]. Following the approach that we have done in this thesis, we could ask if it is possible to express this theory in such a way that only these modes would remain in the theory. Yet, the linear decomposition (4.1) does not allow us to do so.

The action of Yang-Mills theory is given by

$$S = \int d^4x \left[-\frac{1}{2}\mathrm{Tr}(F_{\mu\nu}F^{\mu\nu}) \right]. \tag{4.39}$$

Similarly to massive theory, the A_0 component is not propagating. It satisfies the following constraint:

$$\begin{aligned} -\Delta A_0^a = &-\dot{A}^a_{i,i} - g\epsilon^{abc}\dot{A}^b_i A^c_i - g\epsilon^{abc} A^b_i A^c_{0,i} - g\epsilon^{abc}\partial_i(A^b_i A^c_0) \\ &- g^2\epsilon^{fbc}\epsilon^{fad} A^b_0 A^c_i A^d_i . \end{aligned} \tag{4.40}$$

Decomposing the spatial part of the vector field according to (4.1), the solution of this constraint to the leading order in the coupling constant is given by

$$A_0^a = \dot{\chi}^a + \frac{g\varepsilon^{abc}}{\Delta}\left[\dot{A}_i^b A_i^c + (\Delta\chi^b + 2A_i^b\partial_i)\dot{\chi}^c\right]. \tag{4.41}$$

By substituting this solution back to the action, we arrive at the following Lagrangian density

$$\mathcal{L} = \mathcal{L}_0 + \mathcal{L}_{int}, \qquad \text{where} \tag{4.42}$$

$$\mathcal{L}_0 = -\frac{1}{2}A_i^{Ta}(\Box + m^2)A_i^{Ta} \qquad \text{and}$$

$$\mathcal{L}_{int} = g\varepsilon^{abc}\left[\frac{1}{2}\chi^b\chi_{,i}^c\Box A_i^{Ta} - \dot{\chi}^a\chi_{,i}^c\frac{m^2}{\Delta}\left(\dot{\chi}_{,i}^b\right) + \chi^b A_i^{Tc}\Box A_i^{Ta} + A_{j,i}^{Ta}A_i^{Tb}A_j^{Tc}\right],$$

up to $\mathcal{O}\left(g^2\right)$. This Lagrangian suggest that the longitudinal mode remains on the non-linear level. However, we know that such result is incorrect—only the transverse modes should be pressent in the theory. This suggests that the decomposition that we have used is not good enough for us to derive conclusions on the behaviour of the degrees of freedom.

4.2.2 *The Gauge-Invariant Variables*

In contrast to massive Yang-Mills theory, the massless one has gauge redundancy. The action (4.39) is left unchanged by the following field transformation:

$$A_\mu \to \tilde{A}_\mu = U A_\mu U^\dagger + \frac{i}{g}U_{,\mu}U^\dagger, \tag{4.43}$$

where U is a unitary, space-time dependent matrix.

In the $U(1)$ case, the above transformation becomes simply

$$A_\mu \to \tilde{A}_\mu = A_\mu + \lambda_{,\mu}, \tag{4.44}$$

where λ is now a space-time dependent function. Then, it is easy to see that the transverse modes, defined by the definition

$$A_i = A_i^T + \chi_{,i} \tag{4.45}$$

are gauge-invariant. In other words, under gauge transformation (4.44), they remain the same:

$$A_i^T \to \tilde{A}_i^T = A_i^T. \tag{4.46}$$

In the non-Abelian case, on the other hand, the same does not hold for the transverse modes defined in (4.39). This gives us a hint—*Could a good decomposition be the one in which the transverse modes are gauge-invariant variables?*

The reason to suspect this is provided by both linearised gravity and Maxwell theory. By solving the constraint for A_0, and substituting it in the action of electrodynamics, it is possible to show that only the transverse modes remain in the action [33]. Similarly, the action of linearised gravity can be formulated in terms of the traceless transverse degrees of freedom, $h^T_{ij}T$ [38].

The goal of this part is to show that this is indeed the case. The proper definition of the transverse modes is given by the following decomposition:

$$A_i = \zeta A_i^T \zeta^\dagger + \frac{i}{g}\zeta_{,i}\zeta^\dagger \tag{4.47}$$

where A_i^T are hermitian 2×2 matrices that satisfy

$$A^T_{i,i} = 0 \tag{4.48}$$

and

$$\zeta = e^{-ig\chi}, \tag{4.49}$$

is an $SU(2)$ matrix. Under the gauge transformations, we can find that the components of the above decomposition transform as

$$A_i^T \to \tilde{A}_i^T = A_i^T \qquad \text{and} \qquad \zeta \to \tilde{\zeta} = U\zeta. \tag{4.50}$$

Moreover, we can notice that

$$D_i(\zeta A_i^T \zeta^\dagger) = \partial_i(\zeta A_i^T \zeta^\dagger) + [A_i, \zeta A_i^T \zeta^\dagger] = \zeta A^T_{i,i}\zeta^\dagger = 0. \tag{4.51}$$

In order to confirm that (4.47) properly defines the transverse modes, we will show that now the action (4.39) can be entirely rewritten in terms of the new transverse modes. Thus, as a first step, we will resolve the constraint for the temporal components of the vector field. In matrix formulation, the constraint that they satisfy is given by

$$-\Delta A_0 = -\dot{A}_{i,i} + ig\left\{2[A_i, A_{0,i}] + [A_{i,i}, A_0] + [\dot{A}_i, \dot{A}_i]\right\} + g^2[A_i, [A_0, A_i]] \tag{4.52}$$

For $A_i^T = 0$, the solution of this constraint is given by

$$A_0 = \frac{i}{g}\dot{\zeta}\zeta^\dagger \tag{4.53}$$

However, once we bring back the transverse modes this is no longer the case due to the term $[\dot{A}_i, \dot{A}_i]$. Based on the gauge transformation of the temporal component we can pick the following Ansatz:

$$A_0 = \zeta f\left(A^T\right)\zeta^\dagger + \frac{i}{g}\dot{\zeta}\zeta^\dagger, \tag{4.54}$$

where function $f\left(A^T\right)$ is a function of the transverse modes. Substituting this into (4.52), and taking into account that in general $\zeta \neq 0$, we obtain

$$-\Delta f(A^T) = ig\left\{2[A_i^T, f(A^T)_{,i}] + [\dot{A}_i^T, \dot{A}_i^T]\right\} + g^2[A_i^T, [f(A^T), A_i^T]] \tag{4.55}$$

Assuming that $g \ll 1$, we can resolve this equation perturbatively, and write the full solution as:

$$A_0 = \zeta\frac{1}{\tilde{D}}\left(ig\left[\dot{A}_i^T, A_i^T\right]\right)\zeta^\dagger + \frac{i}{g}\dot{\zeta}\zeta^\dagger \tag{4.56}$$

where

$$\tilde{D} = -\Delta - 2ig\left[A_i^T, \bullet\right] + g^2\left[A_i^T, \left[A_i^T, \bullet\right]\right]. \tag{4.57}$$

Here, $\bullet$ denotes a place in which the expression on which $\frac{1}{\tilde{D}}$ acts upon should be placed when $\tilde{D}$ is perturbatively evaluated.

Now, we should show that by substituting (4.56) the resulting action will not contain ζ, but only the transverse modes. To show it, it is convenient to express A_0 in the following way. Let us denote

$$A_0^T = \frac{1}{\tilde{D}}\left[\dot{A}_i^T, A_i^T\right]. \tag{4.58}$$

Then we can write the vector field as:

$$A_\mu = \zeta A_\mu^T \zeta^\dagger + \frac{i}{g}\zeta_{,\mu}\zeta^\dagger. \tag{4.59}$$

Note that this is exactly the decomposition that was used in [58, 110]. Noticing then that

$$F_{\mu\nu}(A) = \zeta F_{\mu\nu}(A^T)\zeta^\dagger, \tag{4.60}$$

where $F_{\mu\nu}\left(A^T\right)$ is the field strenght tensor that depends only on the transverse modes, the action becomes a function of only transverse modes

$$S = -\frac{1}{2}\int d^4x \mathrm{Tr}\left[F^2(A^T)\right] \tag{4.61}$$

with the same form as (4.39), but with A_0 and A_i replaced by A_0^T and A_i^T respectively.

4.2.3 *The Difference of the Definition of the Degrees of Freedom*

For completeness, let us demonstrate the difference in the definition of the longitudinal and transverse modes in the two formulations. Expanding ζ in the coupling , the spatial component of the vector field becomes

$$\begin{aligned} A_i^a = A_i^{Ta} + \chi_{,i}^a - g\varepsilon^{abc}(A_i^{Tb}\chi^c + \frac{1}{2}\chi_{,i}^b\chi^c) \\ - \frac{g^2}{2}\varepsilon^{fab}\varepsilon^{fcd}(\chi^b A_i^{Tc}\chi^d + \frac{1}{3}\chi^b\chi_{,i}^c\chi^d). \end{aligned} \tag{4.62}$$

This form is particularly encouraging—at $O(g)$ the transverse part of this decomposition matches with the first redefinition of the transverse modes (4.32). By denoting the fields of the linear decomposition as $\tilde{\chi}$ and $\tilde{A}_i^T$, the old and new degrees of freedom are related with the following expressions:

$$\tilde{\chi} = \frac{\partial_i}{\Delta}\left[\zeta A_i^T\zeta^\dagger + \frac{i}{g}\zeta_{,i}\zeta^\dagger\right] \quad \text{and} \quad \tilde{A}_i^T = P_{ij}^T\left[\zeta A_j^T\zeta^\dagger + \frac{i}{g}\zeta_{,j}\zeta^\dagger\right]. \tag{4.63}$$

4.3 The Resolution of the Massless Limit

Previously, we have analyzed both linear and non-linear decompositions in massless Yang-Mills theory. We have seen that in the linear decomposition (4.1), the longitudinal mode remains in the interacting terms. However, we have also shown that these modes shouldn't be present, as massless Yang-Mills theory has only two degrees per color—the transverse modes. With the following decomposition, on the other hand, it is possible to write the theory only in terms of the transverse modes:

$$A_i = \zeta A_i^T\zeta^\dagger + \frac{i}{g}\zeta_{,i}\zeta^\dagger. \tag{4.64}$$

Here, the transverse modes satisfy

$$A_{i,i}^T = 0. \tag{4.65}$$

The longitudinal modes are contained in ζ, a unitary matrix that is given by

$$\zeta = e^{-ig\chi}. \tag{4.66}$$

In this chapter, we will apply this decomposition to the massive Yang-Mills theory. For simplicity, we will now work with matrix formulation of the theory. In this case, the action is given by (2.57). Let us now express it just in terms of the physical degrees of freedom. In order to achieve that, we will first have to express the temporal component in terms of the remaining degrees of freedom. By varying the action (2.57) with respect to the A_0 component, we find the following constraint:

$$\begin{aligned}\left(-\Delta + m^2\right) A_0 = &-\dot{A}_{i,i} + ig[\dot{A}_i, A_i] + ig\left(2\left[A_i, A_{0,i}\right] + \left[A_{i,i}, A_0\right]\right)\\ &+ g^2\left[A_i, [A_0, A_i]\right],\end{aligned} \tag{4.67}$$

where [,] is the commutator, and we have kept A_i intact for simplicity. This constraint is just the matrix version of (4.40). It's solution is given by

$$A_0 = \zeta\frac{1}{D}\left(-\frac{i}{g}m^2\zeta^\dagger\dot{\zeta} + ig\left[\dot{A}_i^T, A_i^T\right]\right)\zeta^\dagger + \frac{i}{g}\dot{\zeta}\zeta^\dagger, \tag{4.68}$$

where

$$\frac{1}{D} = \frac{1}{-\Delta + m^2 - 2ig[A_i^T, \partial_i \bullet\,] + g^2[A_i^T, [A_i^T, \bullet\,]]}. \tag{4.69}$$

The $\bullet$ in this operator denotes the position in which expression that $\frac{1}{D}$ acts upon should be placed, once $\frac{1}{D}$ is perturbatively evaluated.

We can notice that the only modification of massive theory in comparison to the solution of the constraint (4.68) in the massless theory is the first term. If one sets mass to zero, both solutions match, as should be the case.

Substituting (4.68) into (2.57), we arrive at the following Lagrangian density

$$\mathcal{L} = \mathcal{L}_0^T + \mathcal{L}_0^\chi + \mathcal{L}_{int}^T + \mathcal{L}_{int}^{T\chi} \tag{4.70}$$

$$\begin{aligned}\mathcal{L}_0^T &= \mathrm{Tr}\left(\dot{A}_i^T\dot{A}_i^T - A_{i,j}^T A_{i,j}^T - m^2 A_i^T A_i^T\right)\\ \mathcal{L}_0^\chi &= -\frac{m^2}{g^2}\mathrm{Tr}\left[\zeta^\dagger\dot{\zeta}\frac{-\Delta}{-\Delta + m^2}\left(\zeta^\dagger\dot{\zeta}\right) - \zeta^\dagger\zeta_{,i}\zeta^\dagger\zeta_{,i}\right]\\ \mathcal{L}_{int}^{T\chi} &= \frac{2im^2}{g}\mathrm{Tr}\left\{-A_i^T\zeta^\dagger\zeta_{,i} + m^2\zeta^\dagger\dot{\zeta}\frac{1}{D}\left[A_i^T, \frac{1}{-\Delta + m^2}\partial_i(\zeta^\dagger\dot{\zeta})\right]\right\}\end{aligned}$$

$$
\begin{aligned}
&- m^2 \mathrm{Tr}\left\{\zeta^\dagger \dot{\zeta} \frac{1}{D}[\dot{A}_i^T, A_i^T] + [\dot{A}_i^T, A_i^T]\frac{1}{D}(\zeta^\dagger \dot{\zeta})\right\} \\
&+ m^2 \mathrm{Tr}\left\{\zeta^\dagger \dot{\zeta} \frac{1}{D}\left[A_i^T, \left[A_i^T, \frac{1}{-\Delta + m^2}(\zeta^\dagger \dot{\zeta})\right]\right]\right\} \\
\mathcal{L}_{int}^T &= \mathrm{Tr}\left\{-2igA_i^T A_j^T \left(A_{j,i}^T - A_{i,j}^T\right) + g^2 \left[\dot{A}_i^T, A_i^T\right]\frac{1}{D}\left[\dot{A}_j^T, A_j^T\right]\right\} \\
&+ g^2 \mathrm{Tr}\left(A_i^T A_j^T A_i^T A_j^T - A_i^T A_i^T A_j^T A_j^T\right).
\end{aligned}
$$

In comparison to the previous case with linear decomposition, we can notice that now all terms that contain the longitudinal modes disappear if we set $m = 0$. Moreover, this Lagrangian density now has a non-polynomial form, in comparison to the previous polynomial case. It is an analog of the Lagrangian density obtained in [58] in the Stueckelberg form. Nevertheless, the crucial difference with that approach is that here we are evaluating the theory in terms of the degrees of freedom directly.

In the subsequent steps, we will first analyze this theory perturbatively. We can see that in contrast to the previous case with linear decomposition, we can now expand in two parameters—$\frac{1}{D}$ and ζ. After we have determined the scale at which the perturbation theory breaks down, and the modes that enter the strong coupling regime, we will analyze the theory beyond this scale and show that the massless limit of massive Yang-Mills theory is smooth.

4.3.1 The Final Strong Coupling Scale

Let us now develop the perturbation theory. Thus, we will expand both $\frac{1}{D}$ and ζ. We can notice that the operator $\frac{1}{D}$ contains only the transverse modes. From the kinetic term of the transverse modes, given in $\mathcal{L}_0^T$, we can infer that the quantum fluctuations for the transverse modes are given as before, by

$$
\delta A_i^T \sim \frac{1}{L}, \tag{4.71}
$$

for length scales $\frac{1}{L^2} \sim k^2 \gg m^2$. Thus, $\frac{1}{D}$ can be expanded as long as the coupling constant is smaller than unity, which is assumed. The matrix ζ, on the other hand, can only be expanded for the following values of the longitudinal mode

$$
\chi < \frac{1}{g}. \tag{4.72}
$$

Let us assume that this holds. Then, after expanding $\frac{1}{D}$ and ζ, we obtain the following Lagrangian density,

$$\mathcal{L} = \mathcal{L}_0 + \mathcal{L}_{int}, \qquad \text{where} \tag{4.73}$$

$$\mathcal{L}_0 = \mathrm{Tr}\left[-\chi\left(\Box + m^2\right)\frac{-\Delta m^2}{-\Delta + m^2}\chi - A_i^T\left(\Box + m^2\right)A_i^T\right]$$

$$\mathcal{L}_{int} \sim \mathrm{Tr}\left\{igm^4[\chi, \dot{\chi}]\frac{1}{\Delta}(\dot{\chi}) - 2igm^2 A_i^T \chi\chi_{,i} - 2ig\left[\dot{A}_i^T, A_i^T\right]\frac{m^2}{\Delta}(\dot{\chi})\right.$$
$$-2igA_i^T A_j^T(A_{j,i}^T - A_{i,j}^T) + \frac{m^2g^2}{6}\left(\chi_{,\mu}\chi\chi^{,\mu}\chi - \chi_{,\mu}\chi^{,\mu}\chi^2\right)$$
$$\left.-\frac{g^2m^2}{3}A_i^T\left[\chi, \left[\chi_{,i}, \chi\right]\right]\right\},$$

evaluated for length-scales $\frac{1}{L^2} \gg m^2$. Now, we can notice that the kinetic terms agree with those of linear decomposition case. However, the interactions are different. In particular, by setting $m = 0$, all of the longitudinal modes vanish. This also includes the $A^T\chi^2$ combination that has previously caused the strong coupling of the transverse and longitudinal modes. However, again the $O(g^2)$ terms will be important when considering the longitudinal modes. This is apparent if we canonically normalise the longitudinal modes, according to

$$\chi_n = m\sqrt{\frac{-\Delta}{-\Delta + m^2}}\chi. \tag{4.74}$$

Then, we can notice that there is a possibility of a discontinuity in the massless limit within the perturbative approach—some of the interacting terms are singular in mass:

$$\mathcal{L}_{int} \sim \mathrm{Tr}\left\{igm[\chi_n, \dot{\chi}_n]\frac{1}{\Delta}(\dot{\chi}_n) - 2igA_i^T\chi_n\chi_{n,i} - 2ig\left[\dot{A}_i^T, A_i^T\right]\frac{m}{\Delta}(\dot{\chi}_n)\right.$$
$$-2igA_i^T A_j^T(A_{j,i}^T - A_{i,j}^T) + \frac{g^2}{6m^2}\left(\chi_{n,\mu}\chi_n\chi_n^{,\mu}\chi_n - \chi_{n,\mu}\chi_n^{,\mu}\chi_n^2\right)$$
$$\left.-\frac{g^2}{3m}A_i^T\left[\chi_n, \left[\chi_{n,i}, \chi_n\right]\right]\right\}, \tag{4.75}$$

Let us now find the most dominant interaction terms among these. Estimating the derivatives as $\partial_\mu \sim \frac{1}{L}$, we can evaluate them at $O(g)$ as

$$igm[\chi_n, \dot{\chi}_n]\frac{1}{\Delta}(\dot{\chi}_n) \sim gm\chi_n^3, \tag{4.76}$$

$$2igA_i^T \chi_n \chi_{n,i} \sim \frac{g}{L} A_i^T \chi_n^2, \tag{4.77}$$

$$2ig \left[\dot{A}_i^T, A_i^T \right] \frac{m}{\Delta} (\dot{\chi}_n) \sim gm \left(A_i^T \right)^2, \tag{4.78}$$

and

$$2igA_i^T A_j^T (A_{j,i}^T - A_{i,j}^T) \sim \frac{g}{L} \left(A^T \right)^3. \tag{4.79}$$

Taking into account that the minimal level of quantum fluctuations for the normalised longitudinal modes is given by

$$\delta \chi_{nL} \sim \frac{1}{L}, \tag{4.80}$$

we can easily see that the last terms of both rows are of the same order, and more dominant than the first terms of both rows. As long as the coupling constant is smaller than unity, these terms will always be subdominant in comparison to the kinetic terms and hence cannot cause strong coupling.

At $O\left(g^2\right)$, the interaction terms can be evaluated as

$$\begin{aligned} &\frac{g^2}{6m^2} \left(\chi_{n,\mu} \chi_n \chi_n^{;\mu} \chi_n - \chi_{n,\mu} \chi_n^{;\mu} \chi_n^2 \right) \sim \frac{g^2}{(mL)^2} \chi_n^4, \qquad \text{and} \\ &\frac{g^2}{3m} A_i^T \left[\chi_n, \left[\chi_{n,i}, \chi_n \right] \right] \sim \frac{g^2}{mL} A^T \chi_n^3 \end{aligned} \tag{4.81}$$

Clearly, among these terms, the first one is the most dominant. It becomes of the same order as the kinetic term for scales

$$L_{str} \sim \frac{g}{m}. \tag{4.82}$$

This scale agrees with that of the unitarity violation. It is a scale at which the longitudinal modes become strongly coupled. Let us now look at what happens with the transverse modes. For them, the second term in (4.81) is the relevant one. It becomes of the same order as the kinetic term at the scale

$$L \sim \frac{g^2}{m}. \tag{4.83}$$

However, since this scale is smaller than the one at which the longitudinal modes become strongly coupled, it is not trustable.

With our first indicator—the Lagrangian density—we have shown that the longitudinal modes enter the strong coupling regime at the scale that matches with L_u, the scale at which the unitarity seems to be violated. However, given that this scale was inferred from the terms at $O\left(g^2\right)$, we will now confirm it on the level of equations of motion, to make sure that no surprises such as a cancellation of the most important terms arise.

4.3.1.1 The Confirmation of the Strong Coupling Scale

The equations of motion for the longitudinal modes are given by

$$\begin{aligned}\left(\Box+m^2\right)\chi \sim\; & 2ig\left[A_i^T,\chi_{,i}\right]-ig\frac{m^2}{\Delta}\left([\chi,\ddot{\chi}]\right)\\ & +igm^2\left\{2\left[\dot{\chi},\frac{1}{\Delta}(\dot{\chi})\right]+\left[\chi,\frac{1}{\Delta}(\ddot{\chi})\right]\right\}\\ & +2ig\frac{1}{\Delta}\left(\left[\ddot{A}_i^T,A_i^T\right]\right)+\frac{g^2}{6}\left(2\left[\chi_{,\mu},\left[\chi,\chi^{,\mu}\right]\right]+\left[\chi,\left[\chi,\Box\chi\right]\right]\right),\end{aligned} \tag{4.84}$$

while for the transverse modes we obtain:

$$\begin{aligned}\left(\Box+m^2\right)A_k^T = P_{ki}^T\Bigg\{2ig\Bigg(&-m^2\chi\chi_{,i}+m^2\left[A_i^T,\frac{1}{\Delta}(\ddot{\chi})\right]+2m^2\left[\dot{A}_i^T,\frac{1}{\Delta}(\dot{\chi})\right]\\ &+2\left[A_j^T,A_{i,j}^T\right]+\left[A_{j,i}^T,A_j^T\right]\Bigg)-\frac{g^2m^2}{3}\left[\chi,\left[\chi_{,i},\chi\right]\right]\Bigg\}.\end{aligned} \tag{4.85}$$

Let us now develop perturbation theory, by expanding the longitudinal and transverse modes in the powers of the coupling constant:

$$\chi=\chi^{(0)}+\chi^{(1)}+\ldots \qquad \text{and} \qquad A_i^T=A_i^{T(0)}+A_i^{T(1)}+\ldots \tag{4.86}$$

where $\chi^{(0)}$ and $A_i^{T(0)}$ satisfy free equations, given by:

$$\left(\Box+m^2\right)\chi^{(0)}=0 \qquad \text{and} \qquad \left(\Box+m^2\right)A_i^{T(0)}=0. \tag{4.87}$$

Following the previous approach, let us now study the corrections, staring with the transverse modes. The first order corrections for them satisfy the following equation:

$$\begin{aligned}\left(\Box+m^2\right)A_k^{T(1)} \sim 2igP_{ki}^T\Big\{&-m^2\chi^{(0)}\chi_{,i}^{(0)}+2\left[A_j^{T(0)},A_{i,j}^{T(0)}\right]\\ &+\left[A_{j,i}^{T(0)},A_j^{T(0)}\right]\Big\}.\end{aligned} \tag{4.88}$$

Using (4.80) and (4.74), we can find that the minimal amplitude of quantum fluctuations for the original longitudinal modes matches the ones of the linear decomposition, and is given by

$$\delta\chi_L \sim \frac{1}{mL}. \tag{4.89}$$

Taking into account also (4.71), we can easily see that all of the non-linear terms in (4.88) are of the same order, and can be evaluated as

$$2igP^T_{ki}\left\{-m^2\chi^{(0)}\chi^{(0)}_{,i} + 2\left[A^{T(0)}_j, A^{T(0)}_{i,j}\right] + \left[A^{T(0)}_{j,i}, A^{T(0)}_j\right]\right\} \sim \frac{g}{L^3}. \tag{4.90}$$

Therefore, the first order corrections to the transverse modes can be estimated as

$$A^{T(1)}_k \sim \frac{g}{L}. \tag{4.91}$$

We can see that they are always smaller than the kinetic term, in contrast to the case of the linear decomposition, where they have entered a strong coupling regime.

Let's now evaluate the corrections to the longitudinal modes. The first order corrections satisfy the following equation:

$$\left(\Box + m^2\right)\chi^{(1)} \sim 2ig\left[A^T_i, \chi_{,i}\right] \sim \frac{g}{(mL)\,L^2}, \tag{4.92}$$

and thus we can evaluate them as

$$\chi^{(1)} \sim \frac{g}{mL}. \tag{4.93}$$

Since we assume that the coupling is smaller than unity, these terms will also never reach the linear ones. Therefore, to verify the strong coupling scale, we must evaluate the next leading order terms.

The second-order corrections to the longitudinal modes satisfy the following equations:

$$\left(\Box + m^2\right)\chi^{(2)} \sim \frac{g^2}{3}\left[\chi^{(0)}_{,\mu}, \left[\chi^{(0)}, \chi^{(0),\mu}\right]\right] \sim \frac{g^2}{(mL)^3\,L^2}. \tag{4.94}$$

As before, we can evaluate them as

$$\chi^{(2)} \sim \frac{g^2}{(mL)^3}. \tag{4.95}$$

By comparing them with the linear term, $\chi^{(0)} \sim \frac{1}{mL}$, we can see that the two become of the same order at the Vainshtein scale:

$$L_{str} \sim \frac{g}{m}, \tag{4.96}$$

in agreement with the scale found from the Lagrangian density. The second-order corrections to the transverse modes satisfy

$$\left(\Box + m^2\right) A_k^{T(2)} \sim -\frac{g^2 m^2}{3} P_{ik}^T \left\{\left[\chi^{(0)}, \left[\chi_{,i}^{(0)}, \chi^{(0)}\right]\right]\right\} \sim \frac{g^2}{(mL)\, L^3}, \tag{4.97}$$

and thus we can evaluate them as

$$A_k^{T(2)} \sim \frac{g^2}{(mL)\, L}. \tag{4.98}$$

They indicate that the transverse modes enter the strong coupling regime at the length-scale

$$L \sim \frac{g^2}{m}, \tag{4.99}$$

again with agreement with the result inferred from the Lagrangian density. Nevertheless, as this scale is smaller than the Vainshtein scale (4.96), this result is no longer trustable—the perturbation theory in the longitudinal modes that was used to obtain this result is no longer valid at this scale. Our goal is to find the true corrections to the transverse modes. For that, we have to go beyond the perturbation theory, which we do in the next part of this chapter.

4.3.2 *Beyond the Vainshtein Scale*

So far, we have found that the longitudinal modes become strongly coupled at the scale L_{str} that matches with the unitarity violation scale. This scale was induced due to the non-linear terms at $O\left(g^2\right)$ that were purely longitudinal. However, what happens with the remaining higher-order terms. As the more longitudinal modes we have, the greater dominance term has among the remaining ones at each order in the coupling constant, let us consider the interacting terms that contain only the longitudinal modes. Then, the remaining non-linear terms in the Lagrangian can be represented as

$$\mathcal{L}_{HOT} \sim \sum_{n=3}^{\infty} \frac{g^n m^2}{L^2} \chi^{n+2}, \tag{4.100}$$

where HOT stands for higher-order terms. Once we reach the strong coupling scale, these terms become:

$$\mathcal{L}_{HOT} \sim \sum_{n=3}^{\infty} \frac{1}{L_{str}^4}. \tag{4.101}$$

In other words, they become of the same order as the linear and most dominant non-linear term. The reason why this phenomenon occurs can be traced to the longitudinal modes. At the strong coupling scale, the minimal level of quantum fluctuations for the longitudinal modes is given by

$$\delta \chi_{L_{str}} \sim \frac{1}{g}. \tag{4.102}$$

However, to expand ζ, we have assumed that this does not hold. Once the Vainshtein scale is reached, the expansion in the longitudinal modes is no longer valid. Thus, to go beyond the perturbation theory of the longitudinal modes, we should keep ζ intact and evaluate the theory through it. For this, we will return to the initial Lagrangian density (4.70), keeping in mind that the operator $\frac{1}{D}$ can nevertheless still be expanded, and first analyze the new form of the kinetic term of the longitudinal modes.

4.3.2.1 The Non-linear Sigma Model

The kinetic term of the longitudinal modes evaluated at length scales $\frac{1}{L^2} \gg m^2$, with $L < L_{str}$ is given by

$$\mathcal{L}_0^{\chi} \sim \frac{m^2}{g^2} \mathrm{Tr}\left(\zeta^{\dagger}_{,\mu} \zeta^{,\mu}\right). \tag{4.103}$$

This theory is known as the *non-linear sigma model.* In this part, we will analyze this model to determine the minimal amplitude of quantum fluctuations beyond the strong coupling scale. Previously, we have worked with the following parametrization

$$\zeta = e^{-ig\chi}. \tag{4.104}$$

However, now it will be more convenient to work with the matrix elements of ζ. They are given by:

$$\zeta = \begin{pmatrix} \zeta_2^* & \zeta_1 \\ -\zeta_1^* & \zeta_2 \end{pmatrix}. \tag{4.105}$$

As ζ is an $SU(2)$ matrix, the two complex fields, ζ_1 and ζ_2 satisfy the following constraint:

$$|\zeta_1|^2 + |\zeta_2|^2 = 1. \tag{4.106}$$

Let us bring this to an even simpler form. By substituting the following decomposition

$$\zeta_1 = \rho_1 e^{ig\theta_1} \qquad \text{and} \qquad \zeta_2 = \rho_2 e^{ig\theta_2}, \tag{4.107}$$

into (4.106), we obtain

$$|\rho_1|^2 + |\rho_2|^2 = 1. \tag{4.108}$$

Parametrizing these components as

$$\rho_1 = \rho\cos(g\sigma) \qquad \text{and} \qquad \rho_2 = \rho\sin(g\sigma), \tag{4.109}$$

(4.106) implies

$$|\rho|^2 = 1. \tag{4.110}$$

Let us set $\rho = 1$. Then, we obtain the final parametrization of the matrix elements of ζ:

$$\zeta_1 = \cos(g\sigma)e^{ig\theta_1} \qquad \text{and} \qquad \zeta_2 = \sin(g\sigma)e^{ig\theta_2}. \tag{4.111}$$

The Lagrangian density corresponding in terms of these fields becomes

$$\mathcal{L}_0^{\chi} = \frac{1}{2}\left[4m^2\partial_\mu\sigma\partial^\mu\sigma + f^2(\sigma)\partial_\mu\theta_1\partial^\mu\theta_1 + p^2(\sigma)\partial_\mu\theta_2\partial^\mu\theta_2\right]. \tag{4.112}$$

Here,

$$f^2(\sigma) = 4m^2\cos^2(g\sigma) \qquad \text{and} \qquad p^2(\sigma) = 4m^2\sin^2(g\sigma). \tag{4.113}$$

Expanding these functions, we would again arrive at the same situation—the strong coupling of the longitudinal modes, now represented with functions σ, θ_1 and θ_2. However, our goal is to evaluate the theory beyond the Vainshtein scale. Thus, we will have to directly study the Lagrangian density (4.112). In order to find the minimal level of quantum fluctuations, we will canonically normalize the fields according to

$$\sigma_n = 2m\sigma, \qquad \partial_\mu\theta_{1n} = f(\sigma)\partial_\mu\theta_1, \qquad \text{and} \qquad \partial_\mu\theta_{2n} = p(\sigma)\partial_\mu\theta_2. \tag{4.114}$$

Then, the Lagrangian density becomes

$$\mathcal{L}_0^{\chi} = \frac{1}{2}\left[\partial_\mu\sigma_n\partial^\mu\sigma_n + \partial_\mu\theta_{1n}\partial^\mu\theta_{1n} + \partial_\mu\theta_{2n}\partial^\mu\theta_{2n}\right]. \tag{4.115}$$

As before, the minimal level of quantum fluctuations of the normalised fields for scales $k^2 \sim \frac{1}{L^2} \gg m^2$ is given by

$$\delta\sigma_{nL} \sim \frac{1}{L}, \qquad \delta\theta_{n1L} \sim \frac{1}{L}, \qquad \text{and} \qquad \delta\theta_{n2L} \sim \frac{1}{L}. \tag{4.116}$$

Estimating $\sin(g\sigma) \sim \cos(g\sigma) \sim O(1)$, we can evaluate the minimal level of quantum fluctuations of the original fields as:

$$\delta\sigma_L \sim \frac{1}{2g}\frac{k}{k_{str}}, \qquad \delta\theta_{1L} \sim \frac{1}{2g}\frac{k}{k_{str}}, \qquad \text{and} \qquad \delta\theta_{2L} \sim \frac{1}{2g}\frac{k}{k_{str}}. \tag{4.117}$$

Let us now use this result and investigate the massless limit of the massive Yang-Mills theory.

4.3.2.2 The Corrections to the Transverse Modes

Our goal is to find the corrections to the transverse modes due to the longitudinal ones. Within the perturbation theory, these corrections were unreliable—the longitudinal modes entered a strong coupling regime. However, using (4.117), we will now be able to go beyond it.

Our strategy will be the following. First, we will evaluate the most dominant term that might cause a discontinuity using the matrix elements explicitly. Then, we will turn to a quicker analysis, evaluating the theory using ζ. The key that will allow us to perform this analysis is the fact that at the strong coupling scale L_{str}, the perturbation theory in the transverse modes still holds. It only breaks down for the longitudinal ones.

Let us start with the explicit approach. As we are directly considering elements of ζ, we will also evaluate the matrix elements of the transverse modes. These can be written as follows:

$$A_i^T = \begin{pmatrix} -\frac{G_i^T}{2} & -\frac{W_i^{T+}}{\sqrt{2}} \\ -\frac{W_i^{T-}}{\sqrt{2}} & \frac{G_i^T}{2} \end{pmatrix}, \tag{4.118}$$

where G_i^T is the real and $W_i^{\pm}$ complex vector fields. As the transverse modes satisfy

$$A_{i,i}^T = 0, \tag{4.119}$$

the same will hold for its components:

$$G_{i,i}^T = 0 \qquad \text{and} \qquad W_{i,i}^{T\pm} = 0. \tag{4.120}$$

Substituting (4.118) into $\mathcal{L}_0^T$, we obtain

$$\mathcal{L}_0^T = \frac{1}{2}\left[\partial_\mu G_i^T \partial^\mu G_i^T - m^2 G_i^T G_i^T\right] + \partial_\mu W_i^+ \partial^\mu W_i^- - m^2 W_i^+ W_i^- \tag{4.121}$$

The transverse modes are normalized. Thus, the minimal amplitude of quantum fluctuations for the elements of A_i^T is given by:

$$\delta G_L \sim \frac{1}{L} \quad \text{and} \quad \delta W_L^{\pm} \sim \frac{1}{L}. \tag{4.122}$$

Let us now study the interacting terms. The interaction that could be the most problematic one—the source of the apparent discontinuity of the transverse modes within the perturbation theory—is given by:

$$\mathcal{L}_{int}^{T\chi} \supset -\frac{2im^2}{g}\mathrm{Tr}\left(A_i^T \zeta^\dagger \zeta_{,i}\right). \tag{4.123}$$

Substituting (4.118) and (4.111), it becomes:

$$\begin{aligned}
-\frac{2im^2}{g}\mathrm{Tr}\left(A_i^T \zeta^\dagger \zeta_{,i}\right) &= -2im^2\left\{\frac{\sigma_{,i}}{\sqrt{2}}\left[W_i^- e^{ig(\theta_1+\theta_2)} - W_i^{T+} e^{-ig(\theta_1+\theta_2)}\right]\right. \\
&+ i\theta_{1,i} G_i^T \cos^2(g\sigma) + i\theta_{2,i} G_i^T \sin^2(g\sigma) \\
&- i\theta_{1,i}\frac{\cos(g\sigma)\sin(g\sigma)}{\sqrt{2}}\left(W_i^{T-} e^{ig(\theta_1+\theta_2)} + W_i^{T+} e^{-ig(\theta_1+\theta_2)}\right) \\
&\left.+ i\theta_{2,i}\frac{\sin(g\sigma)\cos(g\sigma)}{\sqrt{2}}\left(W_i^{T-} e^{ig(\theta_1+\theta_2)} + W_i^{T+} e^{-ig(\theta_1+\theta_2)}\right)\right\}.
\end{aligned} \tag{4.124}$$

Canonically normalising the longitudinal modes according to (4.114), this term becomes

$$\begin{aligned}
-\frac{2im^2}{g}\mathrm{Tr}\left(A_i^T \zeta^\dagger \zeta_{,i}\right) &\sim -im\left\{\frac{\sigma_{n,i}}{\sqrt{2}}[W_i^{T-} h - W_i^{T+} h^*]\right. \\
&+ iG_i^T\left[\theta_{1n,i}\cos\left(\frac{g\sigma_n}{2m}\right) + \theta_{2n,i}\sin\left(\frac{g\sigma_n}{2m}\right)\right] \\
&\left.+ \frac{i}{\sqrt{2}}\left[\theta_{2n,i}\cos\left(\frac{g\sigma_n}{2m}\right) - \theta_{1n,i}\sin\left(\frac{g\sigma_n}{2m}\right)\right][W_i^{T-} h + W_i^{T+} h^*]\right\},
\end{aligned} \tag{4.125}$$

where

$$h \sim e^{ig(\theta_1+\theta_2)}. \tag{4.126}$$

Estimating

$$\sin(g\sigma) \sim O(1), \qquad \cos(g\sigma) \sim O(1) \qquad \text{and} \qquad h \sim O(1). \tag{4.127}$$

and using (4.116), this term can be evaluated as

$$-\frac{2im^2}{g}\mathrm{Tr}\left(A_i^T \zeta^\dagger \zeta_{,i}\right) \sim g\frac{L}{L_{str}}\frac{1}{L^4}. \tag{4.128}$$

Thus, the divergences that we encountered in the perturbative framework have disappeared. Moreover, as we approach smaller length scales, or equivalently, higher energies, the contribution of the term becomes smaller. As mass vanishes, the strong coupling scale rises to infinity and this term disappears.

To conclude that no problems will arise in the massless limit, we now have to show that the remaining interactions give rise to contributions that are smaller than this interaction. We will analyze this with more efficient methods, but equivalent to the previous consideration. First, we can notice that all of the remaining interactions have the operator $\frac{1}{D}$. However, this operator contains only transverse modes and can always be perturbatively expanded. Thus, we will consider only the leading contribution that is given by

$$\frac{1}{D} \sim \frac{1}{-\Delta}. \tag{4.129}$$

The higher ones will not change the result—they become smaller as the order in the coupling constant increases. Let us further define the following quantity

$$\Omega_0 = \zeta^\dagger \dot{\zeta}, \tag{4.130}$$

as this combination of matrices often appears. Then, the remaining interactions can be expressed as

$$\begin{aligned}\mathcal{L}_{int}^{T\chi} \supset \mathrm{Tr}\Bigg\{&\frac{2im^4}{g}\Omega_0\frac{1}{\Delta}\left[\dot{A}_i^T, \frac{1}{\Delta}\left(\Omega_{0,i}\right)\right] + 2m^2\Omega_0\frac{1}{\Delta}\left[\dot{A}_i^T, A_i^T\right] \\ &-m^4\Omega_0\frac{1}{\Delta}\left[A_i^T, \left[A_i^T, \frac{1}{\Delta}\left(\Omega_0\right)\right]\right]\Bigg\}.\end{aligned} \tag{4.131}$$

In order to evaluate Ω_0, it will be useful to first find the scaling of its matrix elements. Let

$$\Omega_0 = \begin{pmatrix} \psi_0 & \Phi_0 \\ -\Phi_0^* & -\psi_0 \end{pmatrix}, \tag{4.132}$$

with $\mu = 0, 1, 2, 3$. The components of Ω_0 are then given by

$$\begin{aligned}\psi_0 &= -ig(\dot{\theta}_1 \cos^2(g\sigma) + \dot{\theta}_2 \sin^2(g\sigma)) \\ &= -\frac{ig}{2m}(\dot{\theta}_{1n} \cos(g\sigma) + \dot{\theta}_{2n} \sin(g\sigma)) \\ &\sim \frac{g}{mL^2},\end{aligned} \tag{4.133}$$

and

$$\Phi_\mu = g[-\dot{\sigma} + i(\dot{\theta}_1 - \dot{\theta}_2) \cos(g\sigma) \sin(g\sigma)] e^{i\theta_1 + \theta_2} \sim \frac{g}{mL^2}. \tag{4.134}$$

In the second line of the first expression, we have first canonically normalized the fields according to (4.114), and then used the minimal level of quantum fluctuations (4.116). Then, we evaluated the second expression in the same way. As both of the elements of $\Omega_{(0)}$ are of the same order, we can evaluate it as

$$\Omega_0 \sim \frac{g}{mL^2}. \tag{4.135}$$

Nevertheless, this approach requires a dose of care. Due to the presence of $\sin(g\sigma)$ and $\cos(g\sigma)$, the derivatives of $\Omega_{(0)}$ will give rise to the additional factors that are inverse in mass. In the simplest case, with only one derivate we have

$$\partial_\mu \Omega_0 \sim \frac{g^2}{(mL)^2 L^2}. \tag{4.136}$$

Moreover, as we are concerned with the corrections to the transverse modes, the terms $\frac{1}{-\Delta}$ that appear in (4.131) will always act on $\Omega_{(0)}$. Taking this into account, the remaining interactions can be evaluated as

$$\mathrm{Tr}\left\{\frac{2im^4}{g}\Omega_0 \frac{1}{\Delta}\left[\dot{A}_i^T, \frac{1}{\Delta}(\Omega_{0,i})\right]\right\} \sim g^3 \left(\frac{L}{L_{str}}\right)^5, \tag{4.137}$$

$$\mathrm{Tr}\left\{2m^2 \Omega_0 \frac{1}{\Delta}\left[\dot{A}_i^T, A_i^T\right]\right\} \sim g^2 \left(\frac{L}{L_{str}}\right)^2 \tag{4.138}$$

and

$$\mathrm{Tr}\left\{-m^4 \Omega_0 \frac{1}{\Delta}\left[A_i^T, \left[A_i^T, \frac{1}{\Delta}(\Omega_0)\right]\right]\right\} \sim g^4 \left(\frac{L}{L_{str}}\right)^6. \tag{4.139}$$

As we can see, these corrections are smaller than (4.128). Thus, the leading order corrections to the transverse modes due to the longitudinal modes are given by

(4.128). This procedure is equivalent to the analysis on the level of equations of motion. There, the leading order corrections to the transverse modes are therefore given by

$$A_i^{T(1)} \sim -i\frac{m^2}{g}\zeta^\dagger\zeta_{,i} \sim \frac{g}{L^3}\frac{L}{L_{str}}. \tag{4.140}$$

This shows that beyond the strong coupling scale, the massless Yang-Mills theory is recovered up to small corrections that disappear in the massless limit. These corrections are always smaller than the kinetic term. In comparison to the case of the linear decomposition, this result is expected. The transverse modes should not be strongly coupled as indicated by the *asymptotic freedom*—at high energies, the degrees of freedom of massless Yang-Mills theory are weakly coupled [128]. This leads us to conclude that the massless limit of massive Yang-Mills theory is smooth, as it was proposed in [58].

So far we have studied two massive gauge theories—massive Yang-Mills theory and a toy model of self-interacting Proca theory. We have seen that in both theories, the massless limit is smooth. However, we may wonder—*Are there consequences of the smooth massless limit on the relations between massive gauge theories?* In the next chapter, we will extend our investigation of the Vainshtein mechanism and massless limits to another class of gauge theories, and find one of the theoretical consequences—the implication of massless limit on the dualities of massive gauge theories.

Chapter 5
The Massless Limit and Dual Theories

Dual theories—theories that describe different fundamental objects yet share the same physics—hold considerable promise to improve our understanding of physical models. Their application extends to numerous areas and phenomena in string theory, quantum field theory, and condensed matter physics, such as exchanges between weak and strong coupling, magnetic monopoles, effects of bosonization, transformations between two weakly coupled string theories, and even those that connect both string and quantum field theories [69–77].

However, a question of the massless limit adds another perspective to some of the theories that were regarded as dual. In this thesis, we will investigate two of them—a duality of the Proca field and massive anti-symmetric 2-form and the duality between a scalar and massive 3-form. In the following, we will explore these theories by studying their structure and become acquainted with the notion of dualities, beginning with one of the first dual theories discovered: the electromagnetic duality.

5.1 The Electromagnetic Duality

Let us consider free Maxwell theory and ask whether the theory would have the same physics if the electric and magnetic fields exchange places. In this part, following the approach of [68], we will see that this is indeed the case—this theory is self-dual. In other words, its dual theory is again a theory of a massless vector field.

To show this, let us introduce *parent action*—action that can relate two theories of different fields. For the case of Maxwell field, this action is given by

$$S_P = \int d^4x \left[-\frac{1}{4} F_{\mu\nu} F^{\mu\nu} + \frac{1}{2} B_\mu {}^*F^{\mu\nu}_{,\nu} \right], \tag{5.1}$$

A. Hell, *The Massless Limit of Massive Gauge Theories*, Springer Theses,
https://doi.org/10.1007/978-3-032-01090-2_5

with an assumption that the field strength is independent of the vector potential. B_μ is the Lagrange multiplier. By varying with respect to it, we obtain

$$^*F^{\mu\nu}_{,\nu} = 0 \tag{5.2}$$

This is the Bianchi identity, that can be found in standard electrodynamics. It allows the field strength to be expressed through the vector potential

$$F_{\mu\nu} = A_{\nu,\mu} - A_{\mu,\nu}. \tag{5.3}$$

By substituting this back to the parent action, we would simply recover standard electrodynamics. However, as we have initially assumed that the field strength tensor does not depend on the vector potential, by varying (5.1) with respect to it, we find that it satisfies the following relation:

$$F^{\rho\sigma} = \frac{1}{2}\varepsilon^{\rho\sigma\mu\nu}G_{\mu\nu}, \tag{5.4}$$

where

$$G_{\mu\nu} = B_{\nu,\mu} - B_{\mu,\nu}. \tag{5.5}$$

Substituting this equation back into (5.1), we again obtain electromagnetic theory:

$$S_P = -\frac{1}{4}\int d^4 G_{\mu\nu}G^{\mu\nu} \tag{5.6}$$

This procedure is known as the *duality transformation*—it relates the two dual theories. From (5.4), we can notice that this transformation has exchanged the roles of the electric and magnetic fields.

In this part, we have introduced the concept of dual theory and duality transformation. Let us now proceed to the two theories that will be the focus of the following sections—a theory of a Kalb-Ramond field and a 3-form, and examine the dualities of these theories.

5.2 The Kalb-Ramond Theory

The Kalb-Ramond theory is a theory of an anti-symmetric two-form. In the massive case, its action is given by

$$S = \frac{1}{12}\int d^4x \left(H_{\mu\nu\rho}H^{\mu\nu\rho} - 3m^2 B_{\mu\nu}B^{\mu\nu}\right), \tag{5.7}$$

where

$$H_{\mu\nu\rho} = B_{\nu\rho,\mu} + B_{\rho\mu,\nu} + B_{\mu\nu,\rho} \tag{5.8}$$

is the field strength tensor for the Kalb-Ramond field $B_{\mu\nu}$, which is anti-symmetric in its indices.

5.2.1 The Degrees of Freedom of the Kalb-Ramond Field

Let us analyse the degrees of freedom of this theory, following the approach of [129]. First, we will decompose the field into components:

$$\begin{aligned} B_{0i} &= C_i^T + \mu_{,i}, \qquad C_{i,i}^T = 0 \\ B_{ij} &= \varepsilon_{ijk} B_k, \qquad B_i = B_i^T + \phi_{,i}, \qquad B_{i,i}^T = 0 \end{aligned} \tag{5.9}$$

where $i, j, k = 1, 2, 3$. Substituting this into (5.7), we obtain

$$\begin{aligned} S = \frac{1}{2}\int d^4x \Big[& C_i^T(-\Delta + m^2)C_i^T - 2\varepsilon_{ijk} C_i^T \dot{B}_{k,j}^T - m^2 \mu \Delta \mu \\ & + \dot{B}_i^T \dot{B}_i^T - m^2 B_i^T B_i^T - \dot{\phi}\Delta\dot{\phi} - \Delta\phi\Delta\phi + m^2 \phi \Delta \phi \Big]. \end{aligned} \tag{5.10}$$

In contrast to the Proca theory, this theory has two constraints—time derivatives are absent in C_i^T and μ. They satisfy the following constraints:

$$(-\Delta + m^2)C_i^T = \varepsilon_{ijk}\dot{B}_{k,j}^T \qquad \text{and} \qquad \Delta\mu = 0, \tag{5.11}$$

whose solutions are given by:

$$C_i^T = \frac{\varepsilon_{ijk}}{-\Delta + m^2}\dot{B}_{k,j}^T \qquad \text{and} \qquad \mu = 0. \tag{5.12}$$

Substituting them back into the action, we arrive at the following expression:

$$S = -\frac{1}{2}\int d^4x \left[B_i^T(\Box + m^2)\frac{m^2}{-\Delta + m^2} B_i^T + \phi(\Box + m^2)(-\Delta\phi) \right]. \tag{5.13}$$

Here, we will reffer to B_i^T as the transverse modes, and φ as the longitudinal ones. In terms of the canonically normalized variables

$$B^T_{ni} = \sqrt{\frac{m^2}{-\Delta + m^2}} B^T_i \qquad \text{and} \qquad \phi_n = \sqrt{-\Delta}\phi \tag{5.14}$$

the action becomes

$$S = -\frac{1}{2}\int d^4x \left[B^T_{ni}(\Box + m^2)B^T_{ni} + \phi_n(\Box + m^2)\phi_n \right]. \tag{5.15}$$

Thus, we can see that massive Kalb-Ramond theory has three degrees of freedom—the same number as Proca theory. Let us now study the dualization procedure between the two theories.

5.2.2 The Duality of Massive Kalb-Ramond and Proca Fields

The original dualization procedure [78–81] was done on the level of path integral. Without the loss of generality, we will here perform it just by using the action of the theory. Let us start with the action of the Proca theory,

$$S = \int d^4x \left(-\frac{1}{4}F_{\mu\nu}F^{\mu\nu} + \frac{m^2}{2}A_\mu A^\mu \right) \tag{5.16}$$

where $F_{\mu\nu}$ is the field strength tensor for the vector field A_μ given by (5.3). In order to dualize the theory, we can construct the parent action

$$S_P = \int d^4x \left(\frac{1}{4}F_{\mu\nu}F^{\mu\nu} - \frac{1}{2}F_{\mu\nu}\left(A^{\nu,\mu} - A^{\nu,\mu}\right) + \frac{m^2}{2}A_\mu A^\mu \right), \tag{5.17}$$

where the relation (5.3) is not assumed to hold. Nevertheless, we obtain this relation by varying with respect to the field strength tensor, and substituting it into (5.17), the theory becomes Proca theory. By varying with respect to the vector potential we obtain

$$\partial_\mu F^{\mu\nu} + m^2 A^\nu = 0 \tag{5.18}$$

Solving this equation in terms of the vector potential, and substituting it into (5.17), we obtain

$$S_P = \int d^4x \left(-\frac{1}{2m^2}F^{;\mu}_{\mu\nu}F^{\gamma\nu}_{,\gamma} + \frac{1}{4}F_{\mu\nu}F^{\mu\nu} \right). \tag{5.19}$$

Expressing the field strength tensor in terms of its dual,

$$F_{\mu\nu} = m\varepsilon_{\mu\nu\alpha\beta}B^{\alpha\beta}, \tag{5.20}$$

we obtain the action for the Kalb-Ramond field $B_{\mu\nu}$,

$$S = \frac{1}{12}\int d^4x \left(H_{\mu\nu\rho}H^{\mu\nu\rho} - 3m^2 B_{\mu\nu}B^{\mu\nu}\right), \tag{5.21}$$

where $H_{\mu\nu\rho}$ is its field strength tensor. If we would again dualize this action proceeding with the same method, we would obtain Proca theory. Yet, the same duality does not hold if the theories are massless.

5.2.3 The Massless Case

In the case $m = 0$, we can notice that Kalb-Ramond theory does not describe two degrees of freedom as Maxwell theory, the massless counterpart of Proca theory. In this case, only the longitudinal mode remains in the action:

$$S = -\frac{1}{2}\int d^4x\phi(\Box)(-\Delta\phi) \tag{5.22}$$

Thus, we do not even have to try to dualize this theory to a Maxwell theory—the degrees of freedom do not match. Rather, massless Kalb-Ramond theory is dual to a scalar field—this is known as the scalar-tensor duality. Let us show this by following the approach of [68].

In particular, we will show that the action of a massless scalar field

$$S_\phi = \frac{1}{2}\int d^4x\partial_\mu\phi\partial^\mu\phi \tag{5.23}$$

is dual to the massless Kalb-Ramond field:

$$S_{KB} = \frac{1}{12}\int d^4x H_{\mu\nu\rho}H^{\mu\nu\rho} \tag{5.24}$$

To show this, we start with a parent action:

$$S_P = \frac{1}{12}\int d^4x \left(H_{\mu\nu\rho}H^{\mu\nu\rho} + 2\phi\, {}^*H_{\mu}^{;\mu}\right) \tag{5.25}$$

where

$${}^*H^\mu = \varepsilon^{\mu\nu\rho\sigma}H_{\nu\rho\sigma} \tag{5.26}$$

In this action, we do not assume that the relation (5.8) holds—the field strenght and the Kalb-Ramond field are independent fields. ϕ is the Lagrange multiplyer. Varying the parent action with respect to it, we obtain

$$ {}^{*}H_{\mu}^{;\mu} = 0 \tag{5.27} $$

This is the Bianchi identity, present in the Kalb-Ramond theory. It allows us to express the field strenght tensor in terms of the Kalb-Ramond field via (5.8). Substituting this into (5.25) recovers the Kalb-Ramond action. Varying the action with respect to the field strength, we obtain

$$ H_{\mu\nu\rho} = \varepsilon_{\alpha\mu\nu\rho}\phi^{,\alpha}. \tag{5.28} $$

Substituting this relation in (5.25), we recover (5.23). This is the dualization procedure that relates an theory of a massless scalar field with the theory of a massless Kalb-Ramond field.

5.3 The Theory of a Three-Form

Another theory that will interest us is the theory of a 3-form. Its action is given by

$$ S = \int d^4x \left(-\frac{1}{48} W_{\mu\nu\alpha\beta} W^{\mu\nu\alpha\beta} + \frac{m^2}{12} C_{\mu\nu\alpha} C^{\mu\nu\alpha} \right) \tag{5.29} $$

where $W_{\mu\nu\alpha\beta} = C_{\nu\alpha\beta,\mu} - C_{\mu\alpha\beta,\nu} + C_{\beta\mu\nu,\alpha} - C_{\alpha\mu\nu,\beta}$ is the field strength tensor and $C_{\mu\nu\alpha}$ is totally anti-symmetric. Starting from the parent action of the three-forms, and following the approach similar to that in Proca theory it is possible to show that performing the dualization procedure one will arrive to the action of a massive scalar field (see eg. [130, 131]):

$$ S = \frac{1}{2} \int d^4x \left(\phi_{,\mu}\phi^{,\mu} - m^2\phi^2 \right) \tag{5.30} $$

For the upcoming analysis, it will be usefull to study the degrees of freedom of this theory. Therefore, let us separate it into the spatial and temporal part, and decompose the spatial further as

$$ \begin{aligned} C_{0ij} &= \varepsilon_{ijk} \left(C_k^T + \mu_{,k} \right), \qquad C_{k,k}^T = 0 \qquad \text{and} \\ C_{ijk} &= \varepsilon_{ijk}\chi. \end{aligned} \tag{5.31} $$

This leads us to the following Lagrangian density:

$$ \mathcal{L} = \frac{1}{2} \left[m^2 C_i^T C_i^T - \mu \left(-\Delta + m^2 \right) \Delta\mu - 2\dot{\chi}\Delta\mu + \dot{\chi}\dot{\chi} - m^2\chi^2 \right]. \tag{5.32} $$

Integrating out the non-propagating degrees of freedom, μ and C_i^T, (5.32) becomes

$$\mathcal{L} = -\frac{1}{2}\chi\left(\Box + m^2\right)\frac{m^2}{-\Delta + m^2}\chi. \tag{5.33}$$

Substituting the canonically normalised variable

$$\chi_n = \sqrt{\frac{m^2}{-\Delta + m^2}}\chi, \tag{5.34}$$

we obtain

$$\mathcal{L} = -\frac{1}{2}\chi\left(\Box + m^2\right)\frac{m^2}{-\Delta + m^2}\chi. \tag{5.35}$$

5.4 Re-Thinking the Duality

At a first glance, the theories of the massive Kalb-Ramond field and Proca theory seem to be entirely different. One is a theory of an antisymmetric 2-form, while the other describes a massive vector field. Yet, we have seen in the previous parts of this chapter that these theories have the same number of degrees of freedom. Moreover, we have seen that there exists a dualization procedure that can relate the actions of the two theories. Based on these results, there were numerous claims in the literature that the two theories are dual (see eg.[78–94, 96, 105]). In other words, it is said that the two theories have the same physics.

These claims were also extended to Stueckelberg theory, massive p-forms [132–134]and curved spacetime [84]. In the special case of Stueckelberg theory for a $p = 3$ form, the theory was considered with an arbitrary potential [135].

However, if these theories are massless, the duality no longer holds. For example, we have seen that the massless Kalb-Ramond and scalar fields are dual, while the Maxwell field is self-dual [78, 136, 137]. The duality among massless Kalb-Ramond and Maxwell fields, on the other hand, does not exist.

This contrast among the massless theories raises the question about the behavior of the massive ones in the massless limit. So far we have seen that in the presence of self-interactions, a degree of freedom that was absent in massless theory entered a strong coupling regime. This was known in massive gravity and Proca theory, and in this thesis, we have confirmed it also in the case of massive Yang-Mills theory. Yet, the massless Kalb-Ramond theory does not have transverse modes, but a longitudinal one. Thus, if we modify Proca and massive Kalb-Ramond theories by adding to them a self-interaction, it is questionable if their degrees of freedom would behave in the same way. If the theories do indeed have the same physics, this should be the case.

The purpose of the following chapters is to study the behavior of degrees of freedom in the presence of a quartic self-interaction. In particular, we will compare massive Kalb-Ramond and Proca theories, and the theories of a massive scalar field and a 3-form. There, a number of statements of duality between the two have also appeared (see eg. [78–88, 130, 131]). We will find that in all theories but the scalar one, some of the degrees of freedom will become strongly coupled and decouple from the remaining ones. However, this will not be the case for the same types of fields.

5.5 $A\mu$ vs. $B_{\mu\nu}$

Let us now introduce the self-interacting theories. Although we have already studied the specific self-interaction for Proca field, in this part we will study another one, given by

$$S = \int d^4x \left[-\frac{1}{4} F_{\mu\nu} F^{\mu\nu} + \frac{m^2}{2} A_\mu A^\mu + \frac{g^2}{4} \left(A_\mu A^\mu \right)^2 \right]. \tag{5.36}$$

After we analyze the underlying physics of this theory, we will compare it to a self-interacting, massive, Kalb-Ramond field, whose action is given by the following action:

$$S = \int d^4x \left[\frac{1}{12} H_{\mu\nu\rho} H^{\mu\nu\rho} + \frac{m^2}{4} B_{\mu\nu} B^{\mu\nu} + \frac{g^2}{16} \left(B_{\mu\nu} B^{\mu\nu} \right)^2 \right]. \tag{5.37}$$

In both cases, we will assume that the coupling constant is very small, $g^2 \ll 1$. As we have learned in the first chapter, the standard methods suggest that the discontinuity in massive vector theories arises on high energies $k^2 \gg m^2$, as that is when the propagator becomes constant. Therefore we will focus our analysis on this case.

In the following, we will study these theories in a manner similar to the previous approaches to the massive Yang-Mills theory and the toy model. First, we will evaluate the theories perturbatively, and show that the same Vainshtein scale is present in both theories. Nevertheless, we will find that the degrees of freedom behave differently. In Proca theory, the longitudinal modes will enter the strong coupling regime, and remain strongly coupled for scales beyond the Vainshtein radius. The transverse modes, on the other hand, are only weakly coupled and survive the massless limit. By contrast, for the massive Kalb-Ramond field, the opposite is true. There, beyond the Vainshtein scale, the two transverse modes are strongly coupled, while the pseudo-scalar mode is in the weakly coupled regime, surviving the massless limit.

5.5.1 Self-Interacting Proca Theory

Thus far, we have analyzed a self-interacting Proca theory, with a cubic interaction. There, we have found that the perturbative approach indicates a discontinuity within the massless limit due to the longitudinal modes. However, we have also found that such theory has a Vainshtein scale—scale at which the longitudinal modes become strongly coupled, and thereafter decouple from the remaining degrees of freedom.

In this part, we will again analyze the interacting Proca theory, albeit now with quartic self-interaction. The method will nevertheless be similar to the previous approach. As a first step, we will express the action (5.36) only in terms of the propagating degrees of freedom—the transverse and longitudinal modes. Decomposing the vector field according to (2.17), we arrive at the following action:

$$S = \frac{1}{2}\int d^4x \Big\{ A_0\left(-\Delta + m^2\right)A_0 + 2A_0\Delta\dot{\chi} - \left(\dot{\chi}\Delta\dot{\chi} - m^2\chi\Delta\chi\right) \\ + \left(\dot{A}_i^T\dot{A}_i^T - A_{i,j}^T A_{i,j}^T - m^2 A_i^T A_i^T\right) \\ + \frac{g^2}{2}\left[A_0^4 - 2A_0^2\left(A_i + \chi_{,i}\right)^2 + \left(A_i + \chi_{,i}\right)^2\left(A_j + \chi_{,j}\right)^2\right]\Big\}, \tag{5.38}$$

where

$$\left(A_i + \chi_{,i}\right)^2 = \left(A_i + \chi_{,i}\right)\left(A_i + \chi_{,i}\right). \tag{5.39}$$

The A_0 component satisfies the following constraint:

$$\left[-\Delta + m^2 + g^2\left(A_i^T + \chi_{,i}\right)^2\right]A_0 - g^2 A_0^3 = -\Delta\dot{\chi}. \tag{5.40}$$

Estimating $-\Delta \sim k^2 \sim \frac{1}{L^2}$, and assuming that the linear term in this constraint is more dominant than the interacting terms,

$$\frac{1}{L^2} > g^2\left(A_i^T + \chi_{,i}\right)^2 \qquad \text{and} \qquad \frac{1}{L^2} > g^2 A_0^3, \tag{5.41}$$

which we will later verify, we can resolve this constraint as

$$A_0 = \frac{-\Delta}{-\Delta + m^2}\dot{\chi} + \frac{g^2}{-\Delta + m^2}\left[\left(\dot{\chi}^2 - \chi_{,i}\chi_{,i} - 2\chi_{,i}A_i^T - A_i^T A_i^T\right)\dot{\chi}\right] \\ + O\left(g^2(mL)^2\frac{\chi^3}{L}, g^4\frac{\chi^5}{L}\right), \tag{5.42}$$

where $\frac{1}{L}$ denotes a derivative acting on χ. Here, have kept the most important terms. Substituting (5.42) back into the action, we arrive at the following Lagrangian density

$$\mathcal{L} = \mathcal{L}_0 + \mathcal{L}_{int}, \qquad \text{where} \tag{5.43}$$

$$\mathcal{L}_0 = -\frac{1}{2}\chi(\Box + m^2)\frac{m^2(-\Delta)}{-\Delta + m^2}\chi - \frac{1}{2}A_i^T(\Box + m^2)A_i^T \qquad \text{and}$$

$$\mathcal{L}_{int} = \frac{g^2}{4}\left(\chi_{,\mu}\chi^{,\mu}\right)^2 - g^2\chi_{,\mu}\chi^{,\mu}\chi_{,i}A_i^T - \frac{g^2}{2}\chi_{,\mu}\chi^{,\mu}A_i^T A_i^T$$
$$+ g^2\left(\chi_{,i}A_i^T\right)^2 + O\left(g^2m^2\frac{\chi^4}{L^2}, g^2\frac{\chi}{L}\left(A^T\right)^3\right),$$

for scales $\frac{1}{L^2} \gg m^2$. In comparison with the initial formulation given in (5.36), this approach allows us to have a direct view of the behaviour of the transverse and longitudinal modes.

5.5.1.1 The Strong Coupling Scale

Let us now perturbatively evaluate this theory, assuming that the kinetic part of the Lagrangian dominates over the interacting terms. As a first step, we will canonically normalise the longitudinal modes, according to (2.33), thus arriving at

$$\mathcal{L} = \mathcal{L}_0 + \mathcal{L}_{int}, \qquad \text{where now} \tag{5.44}$$

$$\mathcal{L}_0 = -\frac{1}{2}\chi_n(\Box + m^2)\chi_n - \frac{1}{2}A_i^T(\Box + m^2)A_i^T, \qquad \text{and}$$

$$\mathcal{L}_{int} \sim \frac{g^2}{4m^4}\left(\chi_{n,\mu}\chi_n^{;\mu}\right)^2 - \frac{g^2}{m^3}\chi_{n,\mu}\chi_n^{;\mu}\chi_{n,i}A_i^T - \frac{g^2}{2m^2}\chi_{n,\mu}\chi_n^{;\mu}A_i^T A_i^T$$
$$+ \frac{g^2}{m^2}\left(\chi_{n,i}A_i^T\right)^2.$$

We can see that due to the canonical normalization, the inverse factors in mass multiply the interacting terms. Moreover, the more longitudinal modes are present in the interaction, the interaction is more singular in mass. This indicates that this theory has a discontinuity in the massless limit due to the longitudinal modes, similar to the toy model.

At the moment, we are treating this theory perturbatively—assuming that the most dominant terms are the kinetic ones, given in $\mathcal{L}_0$. Nevertheless, there will exist a scale—the Vainshtein scale—at which the perturbation theory will break down due to the non-linear terms. In order to find it, we will now consider the interacting terms specified by $\mathcal{L}_{int}$, and find which among them are the most dominant ones.

Evaluating the derivatives as $\partial_\mu \sim \frac{1}{L}$, and using the minimal amplitude of quantum fluctuations for the longitudinal and transverse modes, given in (3.48), we can evaluate these terms as

$$\frac{g^2}{4m^4}\left(\chi_{n,\mu}\chi_n^{;\mu}\right)^2 \sim \frac{g^2}{(mL)^4L^4}, \tag{5.45}$$

$$\frac{g^2}{m^3}\chi_{n,\mu}\chi_n^{;\mu}\chi_{n,i}A_i^T \sim \frac{g^2}{(mL)^3L^4} \tag{5.46}$$

and

$$\frac{g^2}{2m^2}\chi_{n,\mu}\chi_n^{;\mu}A_i^TA_i^T \sim \frac{g^2}{m^2}\left(\chi_{n,i}A_i^T\right)^2 \sim \frac{g^2}{(mL)^2L^4}. \tag{5.47}$$

The condition that the first term is the most dominant one,

$$mL \ll 1, \tag{5.48}$$

is satisfied automatically—we are evaluating the theory on high energies. This term becomes of the same order as the kinetic at length-scale

$$L_{str} \sim \frac{\sqrt{g}}{m}, \tag{5.49}$$

in agreement with the result of [138], where it was obtained through the introduction of a Stueckelberg field. Therefore, once L_{str} is reached, the perturbation theory breaks down in the longitudinal modes.

The perturbation theory in both longitudinal and transverse modes also suggests that the corrections to the transverse modes due to the longitudinal ones are singular in massless limit. The most important term among these corrections is the following one:

$$\frac{g^2}{m^3}\chi_{n,\mu}\chi_n^{;\mu}\chi_{n,i}A_i^T \sim \frac{g^2}{(mL)^3L^4} \tag{5.50}$$

Moreover, it would suggest that the transverse modes also become strongly coupled, on length-scales

$$L^T \sim \frac{g^{2/3}}{m}. \tag{5.51}$$

Nevertheless, this result is not trustable—the scale L^T is smaller than L_{str}. Thus, the longitudinal modes are strongly coupled before L^T is even reached. We will only be able to obtain the corrections to the transverse modes once we analyze the theory for scales $L \leq L_{str}$. Before we do that, however, let us confirm that L_{str} is implied by the equations of motion following the approach of the toy model. The equations of motion for the longitudinal and transverse modes on scales $\frac{1}{L^2} \gg m^2$ are given by

$$\left(\Box + m^2\right)\chi \sim -\frac{g^2}{m^2}\left(\chi^{,\mu}\chi_{,\mu}\Box\chi + 2\chi_{,\mu}\chi_{,\nu}\chi^{,\mu\nu}\right) + \frac{2g^2}{m^2}\left(\chi_{,i}A_i^T\Box\chi + 2\chi_{,i\mu}\chi^{,\mu}A_i^T + \chi^{,\mu}\chi_{,i}A_{i,\mu}^T\right), \tag{5.52}$$

and

$$\left(\Box + m^2\right)A_k^T \sim -g^2\left(\delta_{ki} - \frac{\partial_k\partial_i}{\Delta}\right)\left(\chi_{,i}\chi_{,\mu}\chi^{\mu} + A_i^T\chi_{,\mu}\chi^{,\mu} - 2\chi_{,i}\chi_{,j}A_j^T\right). \tag{5.53}$$

Here we have again taken into account only the most relevant terms. We can now develop the perturbation theory, expanding the modes in powers of g^2:

$$\begin{aligned} \chi &= \chi^{(0)} + \chi^{(1)} + ..., \qquad \text{and} \\ A_i^T &= A_i^{T(0)} + A_i^{T(1)} + ... \end{aligned} \tag{5.54}$$

$\chi^{(0)}$ and $A_i^{T(0)}$ satisfy the linear equations of motion

$$\left(\Box + m^2\right)\chi^{(0)} = 0 \qquad \text{and} \qquad \left(\Box + m^2\right)A_i^{T(0)} = 0, \tag{5.55}$$

whose solutions are plane waves. The leading correction to the longitudinal mode satisfies

$$\left(\Box + m^2\right)\chi^{(1)} \sim -\frac{2g^2}{m^2}\chi_{,\mu}^{(0)}\chi_{,\nu}^{(0)}\chi^{(0),\mu\nu} + \frac{2g^2}{m^2}\left(2\chi_{,i\mu}^{(0)}\chi^{(0),\mu}A_i^{T(0)} + \chi^{(0),\mu}\chi_{,i}^{(0)}A_{i,\mu}^{T(0)}\right). \tag{5.56}$$

We can evaluate these corrections as

$$\chi^{(1)} \sim \frac{g^2\chi^3}{(mL)^2} \sim \frac{g^2}{(mL)^5}. \tag{5.57}$$

Once they become of the same order as $\chi^{(0)}$, the perturbation theory breaks down. This happens at the Vainshtein radius $L_{str} \sim \frac{\sqrt{g}}{m}$, in agreement with the analysis at the level of Lagrangian density. Using the same procedure, the corrections to the transverse modes due to the longitudinal ones can be estimated as:

$$A_k^{T(1)} \sim \frac{g^2\chi^3}{L} \sim \frac{g^2}{(mL)^3}. \tag{5.58}$$

However, these are not trustworthy due to the strongly coupled longitudinal modes.

5.5.1.2 Beyond the Strong Coupling

For length-scales $L \leq L_{str}$ the most dominant term for the longitudinal modes is given by

$$\mathcal{L}_{\chi int} = \frac{g^2}{4}\left(\chi_{,\mu}\chi^{,\mu}\right)^2. \tag{5.59}$$

With it, we are able to determine the behaviour of the longitudinal modes within the strong coupling regime. It implies a new canonically normalised variable:

$$\chi_n \sim \frac{g}{\sqrt{2}}\frac{\chi^2}{L}, \tag{5.60}$$

with which we can rewrite it as

$$\mathcal{L}_{\chi int} \sim \frac{1}{2}\chi_{n,\mu}\chi_n^{,\mu}. \tag{5.61}$$

The minimal level of quantum fluctuations for the normalised mode is $\delta\chi_n \sim \frac{1}{L}$ similarly to the linear theory. Thus, for the original mode we have

$$\delta_L\chi \sim \frac{1}{\sqrt{g}}. \tag{5.62}$$

We have seen that the perturbation theory in the longitudinal modes has before implied a strong coupling also of the transverse modes. Even though it is no longer valid for the longitudinal modes beyond the Vainshtein scale, it nevertheless holds in the transverse modes. Therefore, let us now find the corrections to the transverse modes due to the longitudinal ones. The Lagrangian density including the most dominant terms relevant for the transverse modes is given by

$$\mathcal{L}_{A^T} \sim -\frac{1}{2}A_i^T(\Box + m^2)A_i^T - g^2\chi_{,\mu}\chi^{,\mu}\chi_{,i}A_i^T + O\left(g^2\frac{\chi^2}{L^2}\left(A^T\right)^2\right). \tag{5.63}$$

Estimating the longitudinal mode as $\chi \sim \frac{1}{\sqrt{g}}$, we can now see that the singularity in mass that we have found for $L > L_{str}$ has disappeared. Rather, the term that contained it will always be smaller than the kinetic term:

$$g^2 \chi_{,\mu}\chi^{,\mu}\chi_{,i}A_i^T \sim \frac{\sqrt{g}}{L^4} \tag{5.64}$$

Therefore, for scales, $L \leq L_{str}$, the corrections to the transverse modes due to the longitudinal ones are given by

$$A_i^{T(1)} \sim \frac{\sqrt{g}}{L}, \tag{5.65}$$

and thus, beyond the Vainshtein radius, the massless theory is recovered up to small corrections. In the massless limit, the strong coupling scale becomes infinite, thus fully recovering the massless theory.

The prerequisite that these results hold is that for all $Lm \ll 1$, the assumptions (5.41) are satisfied. Evaluating the temporal component as $A_0 \sim \dot{\chi}$, and transverse modes and the derivatives as before, both of these assumptions reduce to

$$g^2\chi^2 < 1. \tag{5.66}$$

On the one hand, for $L > L_{str}$, the longitudinal modes can be estimated as $\chi \sim \frac{1}{mL}$. Hence, in this case, the above assumption

$$g^2\left(\frac{L_{str}}{L}\right)^2 < 1 \tag{5.67}$$

is satisfied since $g^2 < 1$. On the other hand, for $L \leq L_{str}$, the longitudinal modes can be evaluated as $\chi \sim \frac{1}{\sqrt{g}}$. Therefore, the above assumption

$$g < 1, \tag{5.68}$$

is again satisfied. Following similar arguments, it is possible to show that the higher-order terms appearing in the interaction Lagrangian, are always subdominant in comparison with the terms given in $\mathcal{L}_{int}$.

The Vainshtein mechanism has also been considered in the case of generalized Proca theory. In this theory, with derivative self-interactions, the longitudinal mode was suppressed due to the Vainshtein mechanism [139].

5.5.2 Self-Interacting Kalb-Ramond Theory

Following the same approach as in the Proca theory, we will now analyze the behavior of the physical degrees of freedom—the transverse and longitudinal modes—of the massive Kalb-Ramond theory in the presence of a self-interaction. As a reminder, its action is given by:

$$S = \int d^4x \left[\frac{1}{12} H_{\mu\nu\rho} H^{\mu\nu\rho} + \frac{m^2}{4} B_{\mu\nu} B^{\mu\nu} + \frac{g^2}{16} \left(B_{\mu\nu} B^{\mu\nu} \right)^2 \right]. \tag{5.69}$$

5.5.2.1 In Search of the Strong Coupling Scale

Our first goal is to perturbatively evaluate the theory and find the scale at which it breaks down. Thus, we will first decompose the Kalb-Ramond field as in the free case (5.9), which will lead us to the following Lagrangain density:

$$\begin{aligned}
\mathcal{L} = & \frac{1}{2} \left[C_i^T (-\Delta + m^2) C_i^T - 2\varepsilon_{ijk} C_i^T \dot{B}_{k,j}^T - m^2 \mu \Delta \mu \right] \\
& + \frac{1}{2} \left(\dot{B}_i^T \dot{B}_i^T - m^2 B_i^T B_i^T - \dot{\phi} \Delta \dot{\phi} - \Delta\phi\Delta\phi + m^2 \phi \Delta \phi \right) \\
& + \frac{g^2}{4} \left(C_i^T + \mu_{,i} \right)^2 \left(C_j^T + \mu_{,j} \right)^2 - \frac{g^2}{2} \left(C_i^T + \mu_{,i} \right)^2 \left(B_j^T + \phi_{,j} \right)^2 \\
& + \frac{g^2}{4} \left(B_i^T + \phi_{,i} \right)^2 \left(B_j^T + \phi_{,j} \right)^2
\end{aligned} \tag{5.70}$$

Similarly to the free case, this theory has two constraints—one for C_i^T and one for μ. By varying the action with respect to them, we obtain respectively:

$$\begin{aligned}
(-\Delta + m^2) C_l^T = & \, \varepsilon_{ljk} \dot{B}_{k,j}^T - g^2 P_{li}^T \Big[C_i^T C_j^T C_j^T + 2 C_i^T C_j^T \mu_{,j} + C_j^T C_j^T \mu_{,i} \\
& + 2 C_j^T \mu_{,i} \mu_{,j} - (C_i^T + \mu_{,i}) \left(B_i^T B_i^T + 2 B_i^T \phi_{,i} + \phi_{,i} \phi_{,i} \right) \\
& + C_i^T \mu_{,j} \mu_{,j} + \mu_{,i} \mu_{,j} \mu_{,j} \Big],
\end{aligned} \tag{5.71}$$

and

$$\begin{aligned}
-m^2 \Delta \mu = & \, g^2 \partial_i \Big[\mu_{,i} \mu_{,j} \mu_{,j} + 2 \mu_{,i} \mu_{,j} C_j^T + \mu_{,j} \mu_{,j} C_i^T + \mu_{,i} C_j^T C_j^T \\
& - (\mu_{,i} + C_i^T) \left(B_j^T B_j^T + 2 B_j^T \phi_{,j} + \phi_{,j} \phi_{,j} \right) \\
& + 2 \mu_{,j} C_j^T C_i^T + C_i^T C_j^T C_j^T \Big].
\end{aligned} \tag{5.72}$$

We can notice that these constraints are now coupled, due to the form of the self-interaction. Let us evaluate them perturbatively. Then, up to $O(g^4)$, the solution to the first constraint is given by

$$
\begin{aligned}
C_i^T &= \varepsilon_{ijk} D\left(\dot{B}_{k,j}^T\right) \\
&+ \frac{g^2 \varepsilon_{ljk} P_{il}^T}{-\Delta + m^2}\left\{D\left(\dot{B}_{k,j}^T\right)\left[\left(B_s^T + \phi_{,s}\right)^2 - \varepsilon_{sab}\varepsilon_{scd} D\left(\dot{B}_{b,a}^T\right) D\left(\dot{B}_{d,c}^T\right)\right]\right\},
\end{aligned}
\tag{5.73}
$$

where

$$
D(B_i) = \frac{1}{-\Delta + m^2}\left[B_i\right], \tag{5.74}
$$

while the solution of the second constraint up to $O(g^4)$ is given by

$$
\begin{aligned}
\mu = \frac{g^2 \varepsilon_{ijk}}{m^2 \Delta}\partial_i &\left\{D\left(\dot{B}_{k,j}^T\right)\left[B_s^T B_s^T + 2B_s^T \phi_{,s} + \phi_{,s}\phi_{,s}\right.\right. \\
&\left.\left. -\varepsilon_{sab}\varepsilon_{scd} D\left(\dot{B}_{b,a}^T\right) D\left(\dot{B}_{d,c}^T\right)\right]\right\}.
\end{aligned}
\tag{5.75}
$$

Substituting them back into the action yields the following Lagrangian density:

$$
\mathcal{L} = \mathcal{L}_0 + \mathcal{L}_{int}, \qquad \text{where} \tag{5.76}
$$

$$
\mathcal{L}_0 = -\frac{1}{2} B_i^T (\Box + m^2)\frac{m^2}{-\Delta + m^2} B_i^T - \frac{1}{2}\phi(\Box + m^2)(-\Delta\phi) \qquad \text{and}
$$

$$
\begin{aligned}
\mathcal{L}_{int} &= \frac{g^2}{4}\left\{\left[\varepsilon_{ijk}\varepsilon_{ils} D\left(\dot{B}_{k,j}^T\right) D\left(\dot{B}_{s,l}^T\right)\right]^2\right. \\
&\left. -2\varepsilon_{ijk}\varepsilon_{ils} D\left(\dot{B}_{k,j}^T\right) D\left(\dot{B}_{s,l}^T\right) B_p^T B_p^T + \left(B_i^T B_i^T\right)^2\right\} \\
&- g^2\left[\varepsilon_{ijk}\varepsilon_{ils} D\left(\dot{B}_{k,j}^T\right) D\left(\dot{B}_{s,l}^T\right) B_p^T \phi_{,p} - B_i^T B_i^T B_j^T \phi_{,j}\right] \\
&- \frac{g^2}{2}\left[\varepsilon_{ijk}\varepsilon_{ils} D\left(\dot{B}_{k,j}^T\right) D\left(\dot{B}_{s,l}^T\right)\phi_{,p}\phi_{,p} - B_i^T B_i^T \phi_{,j}\phi_{,j} - 2\left(B_i^T \phi_{,i}\right)^2\right] \\
&+ O\left(\frac{g^2 B^T \phi^3}{L^3}\right),
\end{aligned}
$$

where we have kept only the most important terms. Let us now determine which among these terms are the most dominant ones, and find the Vainshtein scale of the theory. In order to do this, we first need the find the minimal level of quantum fluctuations of the fields. However, as in the free theory, the $\mathcal{L}_0$ part of the Lagrangian density shows that the fields are not canonically normalised. Canonically normalising them according to (5.14), we can easily find that the mininimal level of quantum fluctuations for the normalised fields for scales $k^2 \sim \frac{1}{L^2} \gg m^2$ is

given by:

$$\delta B^T_{nL} \sim \frac{1}{L} \quad \text{and} \quad \delta\phi_{nL} \sim \frac{1}{L}. \tag{5.77}$$

Therefore, for the original fields we have:

$$\delta B^T_L \sim \frac{1}{mL^2} \quad \text{and} \quad \delta\phi_L \sim O(1). \tag{5.78}$$

This will allow us to compare the interacting terms in (5.76). Estimating the theory for scales $\frac{1}{L^2} \gg m^2$, and evaluating the derivatives as $\partial_\mu \sim \frac{1}{L}$, each row of the interacting Lagrangian can be respectively evaluated as:

$$g^2\left(B^T\right)^4, \quad \frac{g^2\left(B^T\right)^3\phi}{L}, \quad \text{and} \quad \frac{g^2\left(B^T\right)^2\phi^2}{L^2}. \tag{5.79}$$

Taking into account (5.78), we can further evaluate these terms as

$$g^2\left(B^T\right)^4 \sim \frac{g^2}{(mL)^4 L^4}, \tag{5.80}$$

$$\frac{g^2\left(B^T\right)^3\phi}{L} \sim \frac{g^2}{(mL)^3 L^4}, \tag{5.81}$$

and

$$\frac{g^2\left(B^T\right)^2\phi^2}{L^2} \sim \frac{g^2}{(mL)^2 L^4}. \tag{5.82}$$

Clearly, among these, the first term is the most dominant one. However, it does not contain any longitudinal modes, as it were the case in the Proca theory—in Kalb-Ramond theory, the discontinuity in the massless limit is due to the transverse modes. It should be stressed that this is not due to the particular self-interaction that we are considering. Rather, we can trace it to the minimal level of quantum fluctuations for the original transverse modes, that like the longitudinal modes of Proca theory, are singular in mass.

The term that is quartic in the transverse modes becomes of the same order as the linear term at the following Vainshtein radius:

$$L_{str} \sim \frac{\sqrt{g}}{m}. \tag{5.83}$$

This is the same scale that we have encountered in Proca theory—a result that is dependent on self-interaction. The longitudinal modes, on the other hand, enter a

strong coupling regime at a length-scale

$$L_\phi \sim \frac{g^{2/3}}{m} \tag{5.84}$$

due to the second term in (5.82). However, this scale is lower than the strong coupling scale of the transverse modes. Hence, such a result is no longer trustworthy—the perturbation theory breaks down before L_ϕ is reached. In order to find out what is happening to the pseudoscalar, we must go beyond the perturbation theory. Before we do it, however, we will first evaluate the theory at the strong coupling scale, and consider higher-order terms.

5.5.2.2 The Trouble with the Higher-Order Terms

Let us now analyze the higher-order terms. We have seen that the more transverse modes we have in our interaction, the more dominant will be the term. Hence, for every order in the coupling constant, we can only take into account the purely transverse terms, and compare them with the most dominant non-linear term. In this case, the higher-order interactions will take the following form for scales $L \geq L_{str}$:

$$\mathcal{L}_{HOT} \supset \sum_{n=2}^{\infty} \frac{g^{2n+2}}{m^{2n}} \left(B^T\right)^{2n+4} \sim \sum_{n=2}^{\infty} \left(\frac{L_{str}}{L}\right)^{4n+4} \frac{1}{L^4}. \tag{5.85}$$

From here, it is clear that once the strong coupling scale is reached, all of these terms become equally important as the most dominant terms. However, this was not the case in Proca theory. There, the most dominant term for the longitudinal modes has remained the most important one at the strong coupling scale and beyond.

In order to see why such a situation occurs in the Kalb-Ramond theory, we must go back to the constraints. Let us start with the constraint of C_i^T. Previously, we have resolved it as in (5.73). Let us now rewrite this expression considering only the contributions of the transverse modes:

$$C^T \sim B^T + g^2 \left(B^T\right)^3 L^2 \tag{5.86}$$

Using (5.78), we can evaluate the non-linear term as:

$$g^2 \left(B^T\right)^3 L^2 \sim \frac{g^2}{(mL)^3 L}. \tag{5.87}$$

It becomes of the same order as the linear one at the scale

$$L^{(1)} \sim \frac{g}{m}, \tag{5.88}$$

that is lower than the strong coupling scale. Moreover, including the most dominant higher-order contributions, we can evaluate the solution of the constraint as

$$C^T \sim B^T + g^2 \left(B^T\right)^3 L^2 + \sum_{n=2}^{\infty} \frac{g^{2n}}{m^{2n-2}} \left(B^T\right)^{2n+1} L^2. \tag{5.89}$$

However, repeating the comparison with the linear term, it is easy to show that all of the non-linear terms will be subdominant compared to it for scales $L \geq L_{str}$, indicating that we can still use perturbation theory for the C_i^T constraint, and evaluate

$$C_i^T \sim B_i^T. \tag{5.90}$$

The second constraint (5.72), on the other hand, faces an issue. Considering only the most dominant terms, its solution can be evaluated as

$$\mu \sim \sum_{n=1}^{\infty} \left(\frac{g}{m}\right)^{2n} \left(B^T\right)^{2n+1} L \sim \sum_{n=1}^{\infty} \frac{1}{\sqrt{g}} \left(\frac{L_{str}}{L}\right)^{4n+1} \tag{5.91}$$

Once the strong coupling scale is reached, all the terms become equally important and the perturbation theory breaks down. This is the source of the behavior of the higher-order terms in the Lagrangian. Thus, the constraint (5.72) can no longer be evaluated as before once we reach the strong coupling scale and go beyond it.

5.5.2.3 Beyond the Strong Coupling Scale

The previous insight of the breakdown of the perturbative description involving a constraint (5.72) constraint can be evaluated in a more insightful way, by directly inspecting the most important non-linear terms. Along this way, we can evaluate the constraint as follows

$$\frac{m^2 \mu}{L^2} + g^2 \left[\frac{\mu}{L^2}\left(\left(B^T\right)^2 + B^T \phi\right) + \frac{\mu^2}{L^3} B^T + \frac{\mu^3}{L^4}\right] \sim \frac{g^2}{L} B^T \left(\left(B^T\right)^2 + B^T \phi\right), \tag{5.92}$$

where we have replaced $C^T \sim B^T$. Among the terms on the left side, the first one is the most dominant one for the scales $L > L_{str}$. Once the scales $L \sim L_{str}$ are reached, the following nonlinear terms are of the same order of magnitude as the linear one

$$m^2 \mu \sim g^2 \mu \left(B^T\right)^2 \sim g^2 \frac{\mu^2}{L} B^T \sim g^2 \frac{\mu^3}{L^2} \tag{5.93}$$

and the constraint can no longer be perturbatively approximated. For $L < L_{str}$ these nonlinear terms dominate. Hence, we can evaluate the solution of the constraint as

$$\mu \sim O(1)LB^T + O(1)\phi + O\left(\frac{1}{\sqrt{g}}\frac{L}{L_{str}}\right). \tag{5.94}$$

Substituting this to the action, we obtain

$$\mathcal{L} = \mathcal{L}_0 + \mathcal{L}_{int}, \qquad \text{where} \tag{5.95}$$

$$\mathcal{L}_0 = -\frac{1}{2}B_i^T(\Box + m^2)\frac{m^2}{-\Delta + m^2}B_i^T - \frac{1}{2}\phi(\Box + m^2)(-\Delta\phi) \qquad \text{and}$$

$$\mathcal{L}_{int} \sim g^2\left(B^T\right)^4 + g^2\left(B^T\right)^3\frac{\phi}{L} + O\left(g^2\left(B^T\right)^2\frac{\phi^2}{L^2}\right).$$

Here we have replaced the temporal and spatial derivatives with $\frac{1}{L}$. We can infer the canonical normalization for the transverse modes from the first term. The new canonically normalized variable is given by

$$B_n^T \sim gLB^T. \tag{5.96}$$

Thus, for scales $L < L_{str}$, the minimal level of quantum fluctuations for the original field is given by

$$\delta_L B^T \sim \frac{1}{\sqrt{g}L}. \tag{5.97}$$

Let us now evaluate the corrections for the longitudinal modes due to the transverse ones. As before, the most relevant term for the longitudinal modes is the second term of $\mathcal{L}_{int}$. Using (5.97), we can evaluate it as

$$g^2\left(B^T\right)^3\frac{\phi}{L} \sim \frac{\sqrt{g}}{L^4}. \tag{5.98}$$

Thus, beyond the Vainshtein scale, the correction to the original longitudinal mode is now simply $O\left(\sqrt{g}\right)$.

5.5.3 *Comparison of the Theories*

In both of the theories, we have found the same Vainshtein scale. For $L > L_{str}$, we have seen that the divergences in mass were caused by the transverse modes in the Kalb-Ramond theory, whereas in Proca theory, they were caused by the longitudinal modes. At the Vainshtein scale, the longitudinal modes in Proca theory became strongly coupled, and decoupled from the remaining degrees of freedom for the length scales smaller than the Vainshtein scale. In contrast, in Kalb-Ramond theory, the transverse modes became strongly coupled and also decoupled from the longitudinal modes beyond the Vainshtein scale. Thus, for $L < L_{str}$ in the case of Kalb-Ramond the massless theory of longitudinal mode was recovered, while in the case of Proca theory, the massless theory of transverse modes was recovered (Figs. 5.1 and 5.2).

The origin of this discrepancy between the two theories can be traced to the minimal level of quantum fluctuations of the original fields. In the case of Proca theory we have the following situation:

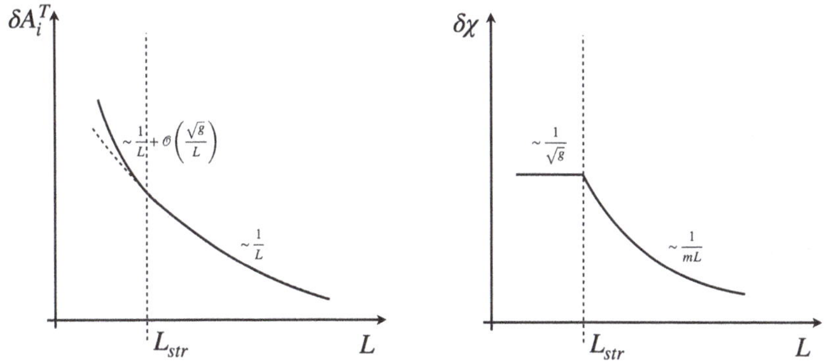

Fig. 5.1 The minimal level of quantum fluctuations of the degrees of freedom in Proca theory with a quartic self-interaction

On the other hand, for the Kalb-Ramond theory we have:

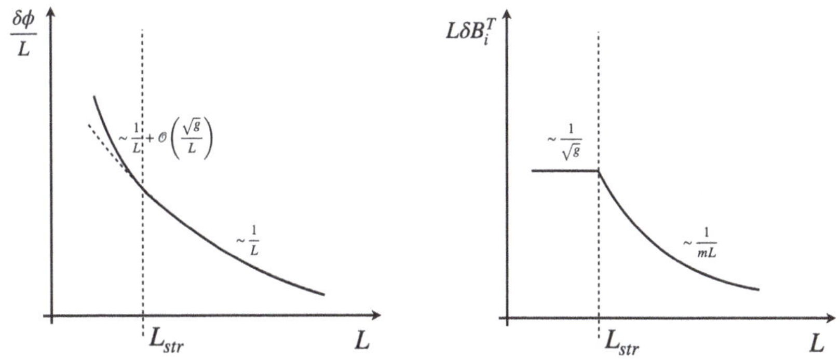

Fig. 5.2 The minimal level of quantum fluctuations of the degrees of freedom in massive Kalb-Ramond theory with a quartic self-interaction

Clearly, the transverse modes of one theory act like the longitudinal modes of the other. This indicates that the two theories do not have the same physics.

Similar to the case that we have considered so far, one could wonder if an analogous situation occurs also for the more general case of massive p-forms. In D dimensions, it is claimed that they are dual to $D - p - 1$-forms. Yet, if massless, they are dual to $D - p - 2$-forms [68, 78, 140]. In the following, we will consider this with an example of duality between a massive 3-form and a scalar field.

5.6 $C_{\mu\nu\rho}$ vs. ϕ

Intuitively, the origin of the difference between Proca and massive Kalb-Ramond fields in the presence of a self-interaction can be found in a different number of degrees of freedom of their massless counterparts. The massless Kalb-Ramond field has only one degree of freedom, while Maxwell theory has two of them. Thus, knowing the Vainshtein mechanism, it was natural to expect that different degrees of freedom would enter the strong coupling regime. As we will see, a similar argument will not only follow for these theories, but also for a theory of a massive self-interacting three form and a scalar field. Let us therefore compare the following two theories:

$$S = \frac{1}{2}\int d^4x \left(\phi_{,\mu}\phi^{,\mu} - m^2\phi^2 - \frac{g^2}{2}\phi^4\right) \tag{5.99}$$

and

$$S = \frac{1}{12}\int d^4x \left[-\frac{1}{4}W_{\mu\nu\alpha\beta}W^{\mu\nu\alpha\beta} + m^2 C_{\mu\nu\alpha}C^{\mu\nu\alpha} + \frac{g^2}{12}\left(C_{\mu\nu\alpha}C^{\mu\nu\alpha}\right)^2\right]. \tag{5.100}$$

Let us now analyse the physical degree of freedom of the self-interacting 3-form. Decomposing it as in the free case (5.31), we arrive at the following Lagrangian density:

$$\begin{aligned} \mathcal{L} = &\frac{1}{2}\left[m^2 C_i^T C_i^T - \mu\left(-\Delta + m^2\right)\Delta\mu - 2\dot{\chi}\Delta\mu + \dot{\chi}\dot{\chi} - m^2\chi^2\right] \\ &+ \frac{g^2}{4}\left[\left(C_i^T + \mu_{,i}\right)^2\left(C_j^T + \mu_{,j}\right)^2 - 2\left(C_i^T + \mu_{,i}\right)^2\chi^2 + \chi^4\right]. \end{aligned} \tag{5.101}$$

Similarly to the case of the Kalb-Ramond field, there are two constraints—one for C_i^T and one for μ—that are given by

$$\left(-\Delta+m^2\right)(-\Delta\mu) =\Delta\dot{\chi}+g^2\partial_i\left[\mu_{,i}\mu_{,j}\mu_{,j}+2\mu_{,i}\mu_{,j}C_j^T+C_i^T\mu_{,j}\mu_{,j}\right.$$
$$\left.+2C_i^TC_j^T\mu_{,j}+\left(\mu_{,i}+C_i^T\right)\left(C_j^TC_j^T-\chi^2\right)\right], \tag{5.102}$$

and

$$m^2C_k^T =-g^2P_{ik}^T\left[C_i^TC_j^TC_j^T+2C_i^TC_j^T\mu_{,j}+C_j^TC_j^T\mu_{,i}+2C_j^T\mu_{,j}\mu_{,i}\right.$$
$$\left.+\left(C_i^J+\mu_{,i}\right)\left(\mu_{,j}\mu_{,j}-\chi^2\right)\right]. \tag{5.103}$$

Up to $O(g^4)$, we can evaluate them respectively as

$$\mu=-D(\dot{\chi})+\frac{g^2}{-\Delta+m^2}\frac{\partial_i}{\Delta}\left\{D(\dot{\chi}_{,i})\left[D(\dot{\chi}_{,j})D(\dot{\chi}_{,j})-\chi^2\right]\right\}, \tag{5.104}$$

and

$$C_j^T=\frac{g^2}{m^2}P_{ij}^T\left\{D(\dot{\chi}_{,i})\left[D(\dot{\chi}_{,j})D(\dot{\chi}_{,j})-\chi^2\right]\right\}, \tag{5.105}$$

where the operator D is given by (5.74). Substituting these solutions back to the Lagrangian density, we obtain:

$$\mathcal{L}=\mathcal{L}_0+\mathcal{L}_{int}, \qquad \text{where} \tag{5.106}$$

$$\mathcal{L}_0=-\frac{1}{2}\chi\left(\Box+m^2\right)\frac{m^2}{-\Delta+m^2}\chi \qquad \text{and}$$
$$\mathcal{L}_{int}=\frac{g^2}{4}\left[D(\dot{\chi}_{,i})D(\dot{\chi}_{,i})\right]^2-\frac{g^2}{2}D(\dot{\chi}_{,i})D(\dot{\chi}_{,i})\chi^2+\frac{g^2}{4}\chi^4+O\left(\frac{g^4\chi^6}{m^2}\right),$$

The situation now is much easier than the theories we have considered before—there is only one field, χ, that we have to compare to the scalar field characterized by the action (5.99). Let us first compare their minimal level of quantum fluctuations. On the one hand, the scalar field is normalized. Hence, its minimal level of quantum fluctuations is given by

$$\delta_L\phi\sim\frac{1}{L} \tag{5.107}$$

for scales $k^2\sim\frac{1}{L^2}\gg m^2$. The field χ, on the other hand, is not. The minimal level of quantum fluctuations of the normalized mode, defined by (5.34), matches that of the scalar field. From it, we can find that the minimal amplitude of the quantum fluctuations for the original mode is given by

$$\delta\chi_L \sim \frac{1}{mL^2}. \tag{5.108}$$

Thus, we can right away see the first difference between the two theories. For the case of a scalar field given in (5.99), there is only one interaction, that can be evaluated as

$$\frac{g^2}{2}\phi^4 \sim \frac{g^2}{L^4}. \tag{5.109}$$

As long as the coupling constant is smaller than unity, it will never become of the same order as the kinetic term. For the pseudoscalar, on the other hand, the non-linear term is singular in mass. The terms at $O(g^2)$ can be evaluated as

$$g^2\chi^4 \sim \frac{g^2}{(mL)^4 L^4}, \tag{5.110}$$

and they become of the same order as the kinetic one at the Vainshtein scale

$$L_{str} \sim \frac{\sqrt{g}}{m}, \tag{5.111}$$

that matches with that of Proca and Kalb-Ramond theory. At the strong coupling scale, the situation is very similar to the case of massive Kalb-Ramond field—the higher order terms that are given by

$$\mathcal{L}_{HOT} \sim \sum_{n=2}^{\infty} \frac{g^{2n}}{(mL)^{2n-2}L^4}(\chi L)^{2n+2} \sim \sum_{n=2}^{\infty}\left(\frac{L_{str}}{L}\right)^{4n}\frac{1}{L^4}, \tag{5.112}$$

all become of the same order as the most dominant term at the strong coupling scale. However, this is now due to the (5.105) constraint, and not the one for μ, as it was previously the case. The solution of this constraint includin the higher order terms can be evaluated as

$$\mu \sim L\chi + g^2(L\chi)^3 + \sum_{n=2}^{\infty}\frac{g^{2n}}{(mL)^{2n-2}}(L\chi)^{2n+1} \tag{5.113}$$

The first non-linear term will become of the same order as the linear one only at the scale

$$L_\mu \sim \frac{g}{m}, \tag{5.114}$$

that is lower than the strong coupling scale. Similarly, the linear term will always be more dominant than the remaining non-linear terms for length scales smaller than the strong coupling scales. The same, however, is not the case for the other constraint. The perturbation theory for it breaks down exactly at L_{str}. Evaluating

$$\mu \sim \chi L \tag{5.115}$$

in the leading order, we can qualitatively represent this constraint as

$$m^2 C^T + g^2 \left[\left(C^T\right)^3 + \left(C^T\right)^2 \chi + C^T \chi^2 \right] \sim g^2 \chi^3 \tag{5.116}$$

At $L \sim L_{str}$, all of the non-linear terms become of the same order as the linear one, as the linear could be evaluated as

$$C^T \sim \frac{g^2}{(mL)^5} \frac{1}{L} \tag{5.117}$$

within the perturbation theory. Thus, perturbation theory for it no longer holds once the strong coupling scale is reached. Beyond it, we can evaluate the leading term to this constraint as

$$C^T \sim \chi, \tag{5.118}$$

leading us to following term:

$$\mathcal{L}_{int} \sim g^2 \chi^4. \tag{5.119}$$

as the most dominant one in the Lagrangian density. Here, we have omitted the temporal and spatial derivatives, that are nevertheless present in contrast to the scalar field.

Below the Vainshtein radius, the longitudinal mode will be in the strongly coupled, with the minimal level of quantum fluctuations given by

$$\delta\chi_L \sim \frac{1}{\sqrt{g}L}. \tag{5.120}$$

We can notice that this is similar to the expressions for the longitudinal modes in Proca theory, and the transverse modes in Kalb-Ramond theory. In contrast, as long as $g \ll 1$, the scalar field ϕ will not become strongly coupled. The origin of the difference between the two theories lies in the minimal quantum fluctuations of the original fields. Thus, following the arguments so far, it is not possible to conclude that these theories are dual.

Chapter 6
Conclusion

In this thesis, we have investigated the massless limit of massive self-interacting gauge theories with mass added *by hand* and its consequences for dual theories. In contrast to the principle of continuity, we have seen that the standard perturbative methods suggest that these theories suffer from a discontinuity in the massless limit.

In the first part , we have explored the massive Yang-Mills theory. It is well established that the massless Yang-Mills theory is unitary and renormalizable [39]. The same is true if the mass of the vector field is generated via the Higgs mechanism [35–37, 40]. Yet, without the Higgs, with mass is just added *by hand*, the resulting theory does not seem to have any of these properties according to the conventional approaches—the perturbative series diverges in the massless limit. In one of the initial studies of this theory, it was suggested that these series might nevertheless be resumed [53]. Yet, the same author has discarded this possibility soon after in [59]—the one-loop corrections were finite but disagreed in a factor of $\frac{1}{2}$ with massless theory.

One of the central aims of this thesis was to show that this discontinuity is just an artifact of the conventional perturbative methods. We have studied this theory through the physical degrees of freedom—the longitudinal and transverse modes—and confirmed that the reason for this discontinuity were the longitudinal modes. Using the methods derived from cosmological perturbation theory, we have shown that these modes, if properly defined, enter a strong coupling regime at the Vainshtein scale, that matches with the scale at which the unitarity violation. This occurs to the non-linear terms that cause a breakdown of the perturbation theory once they become of the same order as the linear terms.

Interestingly, if we would rewrite the massive Yang-Mills theory with the Higgs field in terms of the gauge-invariant variables, as it was done in [38], we would find that the two theories would precisely match if the Higgs would be set to a constant. Otherwise, the theories differ in the presence of a constraint (4.106) due to which we obtain a non-linear sigma model in the theory with mass added *by hand*. Beyond the Vainshtein scale, it is not possible to expand the functions that appear in this model,

A. Hell, *The Massless Limit of Massive Gauge Theories*, Springer Theses,
https://doi.org/10.1007/978-3-032-01090-2_6

and the longitudinal modes remain strongly coupled. However, the transverse modes on these scales are in the weak coupling regime. This has enabled us to find their corrections due to the longitudinal modes. We have found that the massless theory is restored up to very small corrections that become smaller as we approach higher energies. This leads to conclude that the massless limit of this theory is smooth, as initially suggested in [58].

In addition to the massive Yang-Mills theory, we have also studied Proca theory, with two separate interactions—a cubic and a quartic one. In both of them, we have found that the longitudinal mode, a degree of freedom that is not present in Maxwell theory, enters a strong coupling regime due to the non-linear terms. Similar to the massive Yang-Mills theory, beyond the strong coupling scale, this mode remains strongly coupled and decouples from the transverse modes, which in turn remain in the weak coupling regime. Thus the massless theory is restored up small corrections, that disappear in the massless limit.

To date, numerous claims have been made in the literature that Proca theory is dual to the massive Kalb-Ramond theory. These theories have the same number of degrees of freedom. One is longitudinal, and two are transverse. If they indeed have the same physics, we should expect that in the presence of self-interactions, the longitudinal mode of the Kalb-Ramond field becomes strongly coupled. Yet, we have shown that this is not the case.

By studying the Kalb-Ramond theory with quartic self-interaction, we have found that within the perturbation theory, the transverse modes cause a singularity of the perturbative series in the mass. Once the non-linear terms become of the same order as linear ones, at the Vainshtein scale, they become strongly coupled and remain in this regime beyond the Vainshtein scale, decoupling from the longitudinal modes up to small corrections. The longitudinal modes, on the other hand, remain weakly coupled and survive in the massless limit. Even though the Vainshtein scale matches that of the Proca theory with quartic self-interaction, we can notice that the degrees of freedom behave in an entirely different way. Within the perturbation theory, the longitudinal modes of Proca theory cause a discontinuity in the massless limit, while in Kalb-Ramond theory this happens because of the transverse modes. Beyond the Vainshtein scale, the massless theory, which has only two degrees of freedom is recovered in Proca theory. In contrast, in Kalb-Ramond theory, the massless theory that is recovered has only one degree of freedom. This can be traced to the minimal level of quantum fluctuations of the original fields and the absence of the different numbers of them in the corresponding massless theories. Moreover, by comparing a theory of a three-form with quartic self-interaction to the theory of a massive scalar field, we have found a similar discrepancy. We have found that the pseudoscalar, the degree of freedom of the three form becomes strongly coupled, while the massive scalar field does not enter the strong coupling regime at all.

The results of this study thus indicate that the duality of Proca and Kalb-Ramond theories and the duality of a three-form and a massive scalar field is not present, contrary to the numerous claims in the literature, and should be further explored.

Appendix A
The Full Lagrangian and the Feynman Rules for a Toy Model

In this Appendix, we will present the full Lagrangian and the Feynman rules for the toy model, which are relevant when calculating the corrections.

A.1 Lagrangian Density

In terms of the propagating variables, the Lagrangian density has an infinite number of terms. Here, we will present them in the full form up to $\mathcal{O}\left(g^4\right)$ terms, which are relevant for the calculations done in the main chapters. In order to simplify the expressions it will be convenient to define the following quantity:

$$D[\chi] \equiv \frac{-\Delta}{-\Delta + m^2}(\chi), \tag{A.1}$$

where, $\frac{1}{-\Delta+m^2}$ should be understood in terms of the Fourier transform:

$$\frac{1}{|\boldsymbol{k}|^2 + m^2}, \tag{A.2}$$

for modes corresponding to a wave number $\boldsymbol{k}$. The Lagrangian density is given by:

$$\mathcal{L} = \mathcal{L}_0 + \mathcal{L}_g + \mathcal{L}_{g^2} \tag{A.3}$$

$$\begin{aligned}
\mathcal{L}_0 &= \mathcal{L}_{0\chi} + \mathcal{L}_{0A^T} \\
\mathcal{L}_{0\chi} &= -\frac{m^2}{2}\chi\left(\Box + m^2\right)D[\chi] \\
\mathcal{L}_{0A^T} &= -\frac{1}{2}A_i^T\left(\Box + m^2\right)A_i^T
\end{aligned}$$

A. Hell, *The Massless Limit of Massive Gauge Theories*, Springer Theses,
https://doi.org/10.1007/978-3-032-01090-2

$$\mathcal{L}_g = \mathcal{L}_{g\chi} + \mathcal{L}_{g\chi A^T}$$

$$\mathcal{L}_{g\chi} = -\frac{g^2}{4}\left\{(D[\dot{\chi}])^2\Delta\chi + \chi_{,i}\chi_{,i}\left(\Box + m^2\right)D[\chi]\right\}$$

$$\mathcal{L}_{g\chi A^T} = -\frac{g^2}{2}A_i^T\left(\chi_{,i} + \frac{1}{2}A_i^T\right)\left(\Box + m^2\right)D[\chi]$$

$$\mathcal{L}_{g^2} = \mathcal{L}_{g^2 4\chi} + \mathcal{L}_{g^2 4A^T} + \mathcal{L}_{g^2 3\chi A^T} + \mathcal{L}_{g^2 2\chi 2A^T} + \mathcal{L}_{g^2\chi 3A^T}$$

$$\mathcal{L}_{g^2 4\chi} = -\frac{g^4}{8}\left\{(-\Delta\chi)D[\dot{\chi}] + \frac{1}{2}\partial_0(\chi_{,i}\chi_{,i})\right\}$$

$$\frac{1}{-\Delta + m^2}\left\{(-\Delta\chi)D[\dot{\chi}] + \frac{1}{2}\partial_0(\chi_{,j}\chi_{,j})\right\}$$

$$\mathcal{L}_{g^2 4A^T} = -\frac{g^4}{32}\partial_0(A_i^T A_i^T)\frac{1}{-\Delta + m^2}\partial_0(A_j^T A_j^T)$$

$$\mathcal{L}_{g^2 3\chi A^T} = -\frac{g^4}{4}\partial_0(\chi_{,i}A_i^T)\frac{1}{-\Delta + m^2}\left\{(-\Delta\chi)D[\dot{\chi}] + \frac{1}{2}\partial_0(\chi_{,j}\chi_{,j})\right\}$$

$$\mathcal{L}_{g^2 2\chi 2A^T} = -\frac{g^4}{8}\partial_0(A_i^T A_i^T)\frac{1}{-\Delta + m^2}\left\{(-\Delta\chi)D[\dot{\chi}] + \frac{1}{2}\partial_0(\chi_{,j}\chi_{,j})\right\}$$

$$-\frac{g^4}{8}\partial_0(\chi_{,i}A_i^T)\frac{1}{-\Delta + m^2}\partial_0(\chi_{,j}A_j^T)$$

$$\mathcal{L}_{g^2\chi 3A^T} = -\frac{g^4}{8}\partial_0(\chi_{,i}A_i^T)\frac{1}{-\Delta + m^2}\partial_0(A_j^T A_j^T)$$

A.2 The Feynman Rules

When calculating the corrections, we will use the original longitudinal modes instead of the canonically normalized ones. This will not influence the results, but is rather a matter of convenience. Namely, with it, we avoid the square-rooots of the operators (or equivalently, momentum). We will use the following short-hand notation:

$$A(p) = \frac{|\boldsymbol{p}|^2}{|\boldsymbol{p}|^2 + m^2} \tag{A.4}$$

The propagators for the longitudinal and transverse modes are given by (Fig. A.1).

Let us now consider the vertices. In the following rules, all momenta will be outgoing. At $O\left(g^2\right)$, we have the following vertices:

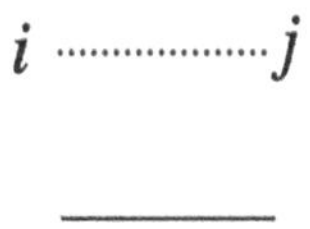

Fig. A.1 Propagators.

$$
\begin{aligned}
V_{3\chi}(k,p,q) &= -\frac{ig^2}{2}\Big[|\boldsymbol{k}|^2 p_0 q_0 A(p)A(q) + (k^2 - m^2)A(k)q_i p_i \\
&\quad + (k \longleftrightarrow p) + ((k \longleftrightarrow q))] \\
V^i_{2\chi}(k_{(i)},p,q) &= -\frac{g^2}{2}\Big[p^i(-q^2+m^2)A(q) + q^i(-p^2+m^2)A(p)\Big] \\
V^{ij}_{\chi}(k,p_{(i)},q_{(j)}) &= \frac{iq^2}{4}(-k^2+m^2)A(k)\eta^{ij},
\end{aligned}
\tag{A.5}
$$

which we have noted by the number of longitudinal mode lines. Furthermore, at $\mathcal{O}\left(g^4\right)$ we will need the following vertices:

$$
\begin{aligned}
V^{ijmn}_{4A}(k,p,q,l) &= -\frac{ig^4}{4}\Bigg[\frac{(k_0+p_0)^2}{|\boldsymbol{k}+\boldsymbol{p}|^2+m^2}\eta^{ij}\eta^{mn} \\
&\quad + (p \longleftrightarrow q)\eta^{im}\eta^{jn} + (p \longleftrightarrow l)\eta^{in}\eta^{jm}\Bigg] \\
V^{ijn}_{3A}(k,p_{(i)},q_{(j)},l_{(n)}) &= \frac{g^4}{4}k_m\Bigg[\frac{(k_0+p_0)^2}{|\boldsymbol{k}+\boldsymbol{p}|^2+m^2}\eta^{mi}\eta^{jn} \\
&\quad + (p \longleftrightarrow q)\eta^{in}\eta^{jm} + (p \longleftrightarrow l)\eta^{ij}\eta^{nm}\Bigg] \\
V^{ij}_{2A}(k,p,q_{(i)},l_{(j)}) &= \frac{ig^4}{8}\Bigg[2\frac{(k_0+p_0)}{|\boldsymbol{k}+\boldsymbol{p}|^2+m^2}\Big(|\boldsymbol{k}|^2A(p)p_0 + |\boldsymbol{p}|^2A(k)k_0 \\
&\quad - (k_0+p_0)k_n p_n\Big)\eta^{ij} + \Bigg(\frac{(k_0+q_0)^2}{|\boldsymbol{k}+\boldsymbol{q}|^2+m^2}k^i p^j \\
&\quad + \frac{(k_0+l_0)^2}{|\boldsymbol{k}+\boldsymbol{l}|^2+m^2}k^j p^i + (p \longleftrightarrow k)\Bigg)\Bigg]
\end{aligned}
\tag{A.6}
$$

$$
\begin{aligned}
V^i_A(k,p,q,l_{(i)}) &= -\frac{g^4}{4}\Bigg[k^i\frac{p_0+q_0}{(\boldsymbol{p}+\boldsymbol{l})^2+m^2}\Big(|\boldsymbol{p}|^2A(q)q_0 + |\boldsymbol{q}|^2A(p)p_0 \\
&\quad - (p_0+q_0)p_n q_n\Big) + (k \longleftrightarrow p) + (k \longleftrightarrow q)\Bigg]
\end{aligned}
\tag{A.7}
$$

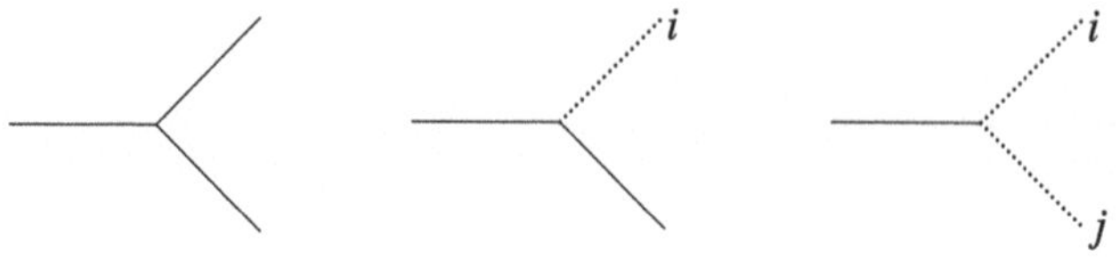

Fig. A.2 3-vertices for the toy model. Reprinted under CC-BY-4.0 license from [REF_NR].

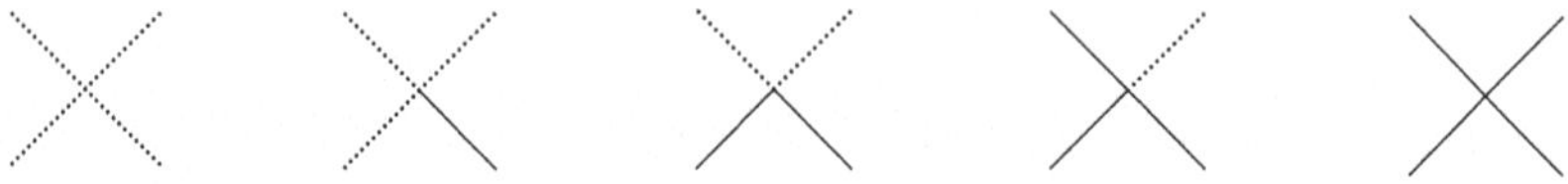

Fig. A.3 4-vertices for the toy model. Reprinted under CC-BY-4.0 license from [REF_NR].

These above vertices are now numbered with the number of transverse lines. The last one in the row contains only the longitudinal modes, so we will not need it for our purposes (Figs. A.2 and A.3).

Appendix B
Lagrangian Density and Feynman Rules in Massive Yang-Mills Theory

In the previous part of the Appendix, we have presented the Lagrangian density and the Feynman rules for the sefl-interacting Proca theory. In this part, we will focus on the massive Yang-Mills theory. In particular, we will present the full Lagrangian density up to $\mathcal{O}\left(g^2\right)$. It will contain both transverse and longitudinal modes. Then, we will show the complete Feynman rules, without the $k^2 \gg m^2$ approximation. In the following, we will use the shorthand notation:

$$D[\chi] \equiv \frac{-\Delta}{-\Delta + m^2}(\chi) \qquad \text{and} \qquad F[\chi] \equiv \frac{\Delta + m^2}{-\Delta + m^2}\chi, \tag{B.1}$$

where $\frac{1}{-\Delta+m^2}$ should be understood in terms of the momentum space. Then, the Lagrangian density up to $\mathcal{O}\left(g^2\right)$ is given by

$$\mathcal{L} = \mathcal{L}_0 + \mathcal{L}_g + \mathcal{L}_{g^2}, \qquad \text{where} \tag{B.2}$$

$$\begin{aligned}
\mathcal{L}_0 &= \mathcal{L}_{0\chi} + \mathcal{L}_{0A^T} \\
\mathcal{L}_{0\chi} &= -\frac{m^2}{2}\chi^a\left(\Box + m^2\right)D[\chi^a] \\
\mathcal{L}_{0A^T} &= -\frac{1}{2}A_i^{Ta}\left(\Box + m^2\right)A_i^{Ta} \\
\mathcal{L}_g &= \mathcal{L}_{g\chi} + \mathcal{L}_{gA^T} + \mathcal{L}_{g2\chi A^T} + \mathcal{L}_{g\chi 2A^T} \\
\mathcal{L}_{g\chi} &= gm^2\varepsilon^{abc}D[\dot{\chi}^a]\frac{1}{-\Delta + m^2}\left(\dot{\chi}^b_{,i}\right)\chi^c_{,i} \\
\mathcal{L}_{gA^T} &= g\varepsilon^{abc}A^{Ta}_{j,i}A_i^{Tb}A_j^{Tc}
\end{aligned}$$

A. Hell, *The Massless Limit of Massive Gauge Theories*, Springer Theses,
https://doi.org/10.1007/978-3-032-01090-2

$$
\begin{aligned}
\mathcal{L}_{g\chi} &= gm^2\varepsilon^{abc}D[\dot{\chi}^a]\frac{1}{-\Delta+m^2}\left(\dot{\chi}^b_{,i}\right)\chi^c_{,i}\\
\mathcal{L}_{gA^T} &= g\varepsilon^{abc}A^{Ta}_{j,i}A^{Tb}_iA^{Tc}_j\\
\mathcal{L}_{g2\chi A^T} &= g\varepsilon^{abc}\left\{D[\dot{\chi}^a]\left[\frac{m^2}{-\Delta+m^2}\left(\dot{\chi}^b_{,i}\right)A^{Tc}_i+\dot{A}^{Tb}_i\chi^c_{,i}\right]+A^{Ta}_{j,i}\chi^b_{,i}\chi^c_{,j}\right\}\\
\mathcal{L}_{g\chi 2A^T} &= g\varepsilon^{abc}\left(D[\dot{\chi}^a]\dot{A}^{Tb}_iA^{Tc}_i+A^{Ta}_{j,i}\chi^b_{,i}A^{Tc}_j\right)\\
\mathcal{L}^2_g &= \mathcal{L}_{g^2\chi}+\mathcal{L}_{g^2A^T}+\mathcal{L}_{g^23\chi A^T}+\mathcal{L}_{g^22\chi 2A^T}+\mathcal{L}_{g^2\chi 3A^T}\\
\mathcal{L}_{g^2\chi} &= -\frac{1}{2}g^2\varepsilon^{fab}\varepsilon^{fcd}\left\{\chi^b_{,i}F[\dot{\chi}^a_{,i}]-\Delta(\chi^b)D[\dot{\chi}^a]\right\}\frac{1}{-\Delta+m^2}\left\{\chi^d_{,j}F[\dot{\chi}^c_{,j}]\right.\\
&\quad\left.-\Delta(\chi^d)D[\dot{\chi}^c]\right\}+\frac{1}{2}g^2\varepsilon^{fab}\varepsilon^{fcd}\left[D[\dot{\chi}^a]\chi^b_{,i}D[\dot{\chi}^c]\chi^d_{,i}\right.\\
&\quad\left.-\frac{1}{2}\chi^a_{,i}\chi^b_{,j}\chi^c_{,i}\chi^d_{,j}\right]\\
\mathcal{L}_{g^2A^T} &= -\frac{1}{2}g^2\varepsilon^{fab}\varepsilon^{fcd}\left\{\left[\dot{A}^{Ta}_iA^{Tb}_i\right]\frac{1}{-\Delta+m^2}\left[\dot{A}^{Tc}_jA^{Td}_j\right]\right.\\
&\quad\left.+\frac{1}{2}A^{Ta}_iA^{Tb}_jA^{Tc}_iA^{Td}_j\right\}\\
\mathcal{L}_{g^23\chi A^T} &= -g^2\varepsilon^{fab}\varepsilon^{fcd}\left\{\dot{A}^{Ta}_i\chi^b_{,i}+A^{Tb}_iF[\dot{\chi}^a_{,i}]\right\}\frac{1}{-\Delta+m^2}\left\{\chi^d_{,j}F[\dot{\chi}^c_{,j}]\right.\\
&\quad\left.-\Delta(\chi^d)D[\dot{\chi}^c]\right\}+g^2\varepsilon^{fab}\varepsilon^{fcd}\left\{D[\dot{\chi}^a]A^{Tb}_iD[\dot{\chi}^c]\chi^d_{,i}-\chi^a_{,j}A^{Tb}_i\chi^d_{,i}\chi^c_{,j}\right\}\\
\mathcal{L}_{g^22\chi 2A^T} &= -\frac{1}{2}g^2\varepsilon^{fab}\varepsilon^{fcd}\left\{2\left[\dot{A}^{Ta}_iA^{Tb}_i\right]\frac{1}{-\Delta+m^2}\left[\dot{A}^{Tc}_j\chi^d_{,j}+A^{Td}_jF[\dot{\chi}^c_{,j}]\right]\right.\\
&\quad\left.+\left[\dot{A}^{Ta}_i\chi^b_{,i}+A^{Tb}_iF[\dot{\chi}^a_{,i}]\right]\frac{1}{-\Delta+m^2}\left[\dot{A}^{Tc}_j\chi^d_{,j}+A^{Td}_jF[\dot{\chi}^c_{,j}]\right]\right\}\\
&\quad-\frac{1}{2}g^2\varepsilon^{fab}\varepsilon^{fcd}\left\{-D[\dot{\chi}^a]A^{Tb}_iA^{Td}_iD[\dot{\chi}^c]+A^{Ta}_iA^{Tb}_j\chi^c_{,i}\chi^d_{,j}\right.\\
&\quad\left.+A^{Ta}_iA^{Tc}_i\chi^b_{,j}\chi^d_{,j}+A^{Ta}_iA^{Td}_j\chi^b_{,j}\chi^c_{,i}\right\}\\
\mathcal{L}_{g^2\chi 3A^T} &= -g^2\varepsilon^{fab}\varepsilon^{fcd}\left\{\left[\dot{A}^{Ta}_iA^{Tb}_i\right]\frac{1}{-\Delta+m^2}\left[\dot{A}^{Tc}_j\chi^d_{,j}+A^{Td}_jF[\dot{\chi}^c_{,j}]\right]\right.\\
&\quad\left.+A^{Ta}_iA^{Tb}_jA^{Tc}_i\chi^d_{,j}\right\}.
\end{aligned}
$$

Fig. B.1 The propagators for the massive Yang-Mills theory

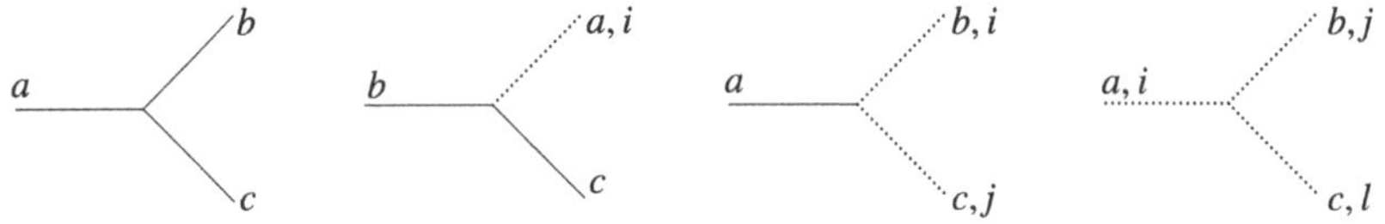

Fig. B.2 3-vertices. Reprinted under CC-BY-4.0 license from [REF_NR]. ©2022, The Author(s)

The Feynman rules involving the original longitudinal modes are given in the following. The propagators for the longitudinal and transverse modes are each given in momentum space by (Fig. B.1):

$$\Delta_{\chi}^{ab}(k) = \frac{|\boldsymbol{k}|^2 + m^2}{m^2|\boldsymbol{k}|^2} \frac{i\delta^{ab}}{k^2 - m^2}$$

$$\Delta_{ij}^{Tab}(k) = \left(\delta_{ij} - \frac{k_i k_j}{|\boldsymbol{k}|^2}\right) \frac{i\delta^{ab}}{k^2 - m^2}$$

At $O(g)$ we have the following vertices, numbered according to the increasing number of longitudinal modes, and with all moments outgoing (Fig. B.2):

$$\begin{aligned}
V_{3\chi}^{abc}(k, p, q) = & \, igm^2\varepsilon^{abc}\left[\frac{|\boldsymbol{k}|^2 k_0 p_i q_i}{|\boldsymbol{k}|^2 + m^2}\left(\frac{p_0}{|\boldsymbol{p}|^2 + m^2} - \frac{q_0}{|\boldsymbol{q}|^2 + m^2}\right)\right. \\
& + \frac{|\boldsymbol{p}|^2 p_0 k_i q_i}{|\boldsymbol{p}|^2 + m^2}\left(\frac{q_0}{|\boldsymbol{q}|^2 + m^2} - \frac{k_0}{|\boldsymbol{k}|^2 + m^2}\right) + \\
& \left. + \frac{|\boldsymbol{q}|^2 q_0 k_i p_i}{|\boldsymbol{q}|^2 + m^2}\left(\frac{k_0}{|\boldsymbol{k}|^2 + m^2} - \frac{p_0}{|\boldsymbol{p}|^2 + m^2}\right)\right] \\
V_{i,2\chi}^{abc}(k_{(i)}, p, q) = & - g\varepsilon^{abc}\left[k_\mu\left(q^\mu p_i - p^\mu q_i\right) + m^2 k_0\left(\frac{p_0 q_i}{|\boldsymbol{p}|^2 + m^2}\right.\right. \\
& \left.\left. - \frac{q_0 p_i}{|\boldsymbol{q}|^2 + m^2}\right) + \frac{m^2 p_0 q_0\left(|\boldsymbol{p}|^2 q_i - |\boldsymbol{q}|^2 p_i\right)}{(|\boldsymbol{p}|^2 + m^2)(|\boldsymbol{q}|^2 + m^2)}\right] \\
V_{ij,\chi}^{abc}(k, p_{(i)}, q_{(j)}) = & - ig\varepsilon^{abc}\delta_{ij}\left[\frac{|\boldsymbol{k}|^2 k_0}{|\boldsymbol{k}|^2 + m^2}(p_0 - q_0) - k_l(p - q)_l\right] \\
V_{ijl}^{abc}(k, p, q) = & g\varepsilon^{abc}\left[\delta_{il}(k - q)_j + \delta_{ij}(p - k)_l + \delta_{jl}(q - p)_i\right]
\end{aligned}$$

Appendix C
Massless Yang-Mills Theory and the Radiation Gauge

Throughout this book, we were mostly focused on the propagating degrees of freedom of the massive Yang-Mills theory—the transverse and longitudinal modes. At the same time, it is important to compare those results to the massless case, in which only the transverse modes are present. One example when such computation is important is the propagator. In the massive case, we can see that the propagator of the transverse modes agrees with the propagator of the massless theory in the radiation gauge, once one takes the massless limit. Therefore, in order to make the comparison simple, we will study the massless theory in the radiation gauge, and consider only the propagating degrees of freedom, the transverse modes. The starting action is given by: The action of this theory is given by

$$S = -\frac{1}{2}\int d^4x \mathrm{Tr}\left(F_{\mu\nu}F^{\mu\nu}\right). \tag{C.1}$$

First, let us decompose the vector field in terms of temporal and spatial part, A_0^a and A_i^a with $a = 1, 2, 3$. Then, the action becomes

$$\begin{aligned} S = \int d^4x \Bigg\{ &\frac{1}{2}\left[A_0^a(-\Delta)A_0^a + 2A_0^a\dot{A}_{i,i}^a\right] + \frac{1}{2}(\dot{A}_i^a\dot{A}_i^a + A_{j,i}^a A_{i,j}^a - A_{j,i}^a A_{j,i}^a) \\ &+ g\epsilon^{abc}(-A_0^b A_i^c \dot{A}_i^a - A_i^b A_0^c A_{0,i}^a + A_i^b A_j^c A_{j,i}^a) \\ &- \frac{g^2}{4}\epsilon^{abc}\epsilon^{aed}(-2A_0^b A_i^c A_0^e A_i^d + A_i^b A_j^c A_i^e A_j^d)\Bigg\}. \end{aligned} \tag{C.2}$$

As a next step, let us choose the radiation gauge condition:

$$A_{i,i}^a = 0.$$

A. Hell, *The Massless Limit of Massive Gauge Theories*, Springer Theses,
https://doi.org/10.1007/978-3-032-01090-2

Due to the absence of derivatives with respect to time, the temporal component is not propagating as in the massive theory. Therefore, the theory has only two degrees of freedom. Let us now find and solve the constraint of the non-propagating component, following the procedure as in the massive case. The A_0 component satisfies the following system of constraints (one for each $a = 1, 2, 3$):

$$\begin{aligned} -\Delta A_0^a &= -\dot{A}_{i,i}^a - g\epsilon^{abc}\dot{A}_i^b A_i^c - g\epsilon^{abc}A_i^b A_{0,i}^c \\ &\quad - g\epsilon^{abc}\partial_i(A_i^b A_0^c) - g^2\epsilon^{fbc}\epsilon^{fad}A_0^b A_i^c A_i^d \end{aligned} \tag{C.3}$$

Up to $O(g)$, the solution of this system can be can be evaluated as

$$\begin{aligned} A_0^a &= -g\varepsilon^{abc}\frac{1}{-\Delta}\left[\dot{A}_i^b A_i^c\right] + 2g^2\varepsilon^{fab}\varepsilon^{fcd}\frac{1}{-\Delta}\left[(A_i^b\partial_i)\frac{1}{-\Delta}\left(\dot{A}_j^c A_j^d\right)\right] \\ &\quad + O\left(g^3\right), \end{aligned} \tag{C.4}$$

and therefore, we obtain the following Lagrangian density:

$$\begin{aligned} \mathcal{L} &= \mathcal{L}_0 + \mathcal{L}_{int}, \qquad \text{where} \\ \mathcal{L}_0 &= -\frac{1}{2}A_i^{Ta}(\Box)A_i^{Ta}, \qquad \text{and} \\ \mathcal{L}_{int} &= g\varepsilon^{abc}A_{j,i}^{Ta}A_i^{Tb}A_j^{Tc} - \frac{1}{2}g^2\varepsilon^{fab}\varepsilon^{fcd}\left\{\left[\dot{A}_i^{Ta}A_i^{Tb}\right]\frac{1}{-\Delta}\left[\dot{A}_j^{Tc}A_j^{Td}\right]\right. \\ &\quad \left. + \frac{1}{2}A_i^{Ta}A_j^{Tb}A_i^{Tc}A_j^{Td}\right\}. \end{aligned} \tag{C.5}$$

We can see that the propagator matches to the one of transverse modes in the massive case if mass set to zero:

$$\Delta_{ij}^{Tab}(k) = \left(\delta_{ij} - \frac{k_i k_j}{|\boldsymbol{k}|^2}\right)\frac{i\delta^{ab}}{k^2},$$

and, the vertex at $O(g)$ exactly coincides with the vertex $V_{ijl}^{abc}(k, p, q)$ of the massive theory.

References

1. A.A. Starobinsky, A new type of isotropic cosmological models without singularity Phys. Lett. B **91**, 99–102 (1980). https://doi.org/10.1016/0370-2693(80)90670-X
2. K. Sato, First order phase transition of a vacuum and expansion of the universe. Mon. Not. Roy. Astron. Soc. **195**, 467–479 (1981). NORDITA-80-29
3. A.H. Guth, The inflationary universe: a possible solution to the horizon and flatness problems. Phys. Rev. D **23**, 347–356 (1981). https://doi.org/10.1103/PhysRevD.23.347
4. A.D. Linde, A new inflationary universe scenario: a possible solution of the horizon, flatness, homogeneity, isotropy and primordial monopole problems. Phys. Lett. B **108**, 389–393 (1982). https://doi.org/10.1016/0370-2693(82)91219-9
5. R. Brout, F. Englert, E. Gunzig, The causal universe. Gen. Rel. Grav. **10**, 1–6 (1979). https://doi.org/10.1007/BF00757018
6. A. Albrecht, P.J. Steinhardt, Cosmology for grand unified theories with radiatively induced symmetry breaking. Phys. Rev. Lett. **48**, 1220–1223 (1982). https://doi.org/10.1103/PhysRevLett.48.1220
7. A.D. Linde, Chaotic inflation. Phys. Lett. B **129**, 177–181 (1983). https://doi.org/10.1016/0370-2693(83)90837-7
8. G.V. Chibisov, V.F. Mukhanov, Galaxy formation and phonons. Mon. Not. Roy. Astron. Soc. **200**, 535–550 (1982)
9. V.F. Mukhanov, G.V. Chibisov, Quantum fluctuations and a nonsingular universe. JETP Lett. **33**, 532–535 (1981)
10. A.A. Starobinsky, Spectrum of relict gravitational radiation and the early state of the universe. JETP Lett. **30**, 682–685 (1979)
11. H. Kodama, M. Sasaki, Cosmological perturbation theory. Prog. Theor. Phys. Suppl. **78**, 1–166 (1984). https://doi.org/10.1143/PTPS.78.1
12. V.F. Mukhanov, Gravitational instability of the universe filled with a scalar field. JETP Lett. **41**, 493–496 (1985)
13. M. Sasaki, Large scale quantum fluctuations in the inflationary universe. Prog. Theor. Phys. **76**, 1036 (1986). https://doi.org/10.1143/PTP.76.1036
14. P.A.R. Ade, et al. [Planck], Planck 2015 results. XIII. Cosmological parameters. Astron. Astrophys. **594**, A13 (2016). https://doi.org/10.1051/0004-6361/201525830. [arXiv:1502.01589 [astro-ph.CO]]
15. A.G. Riess, et al. [Supernova Search Team], Observational evidence from supernovae for an accelerating universe and a cosmological constant. Astron. J. **116**, 1009–1038 (1998). https://doi.org/10.1086/300499. [arXiv:astro-ph/9805201 [astro-ph]]

A. Hell, *The Massless Limit of Massive Gauge Theories*, Springer Theses,
https://doi.org/10.1007/978-3-032-01090-2

16. S. Perlmutter, et al. [Supernova Cosmology Project]. Measurements of Ω and Λ from 42 high redshift supernovae. Astrophys. J. **517**, 565–586 (1999). https://doi.org/10.1086/307221. [arXiv:astro-ph/9812133 [astro-ph]]
17. P.J.E. Peebles, B. Ratra, Cosmology with a time variable cosmological constant. Astrophys. J. Lett. **325**, L17 (1988). https://doi.org/10.1086/185100
18. Y. Fujii, Origin of the gravitational constant and particle masses in scale invariant scalar - tensor theory. Phys. Rev. D **26**, 2580 (1982). https://doi.org/10.1103/PhysRevD.26.2580
19. L.H. Ford, Cosmological constant damping by unstable scalar fields. Phys. Rev. D **35**, 2339 (1987). https://doi.org/10.1103/PhysRevD.35.2339
20. C. Wetterich, Cosmology and the fate of dilatation symmetry. Nucl. Phys. B **302**, 668–696 (1988). https://doi.org/10.1016/0550-3213(88)90193-9. [arXiv:1711.03844 [hep-th]]
21. B. Ratra, P.J.E. Peebles, Cosmological consequences of a rolling homogeneous scalar field. Phys. Rev. D **37**, 3406 (1988). https://doi.org/10.1103/PhysRevD.37.3406
22. T. Chiba, N. Sugiyama, T. Nakamura, Cosmology with x matter. Mon. Not. Roy. Astron. Soc. **289**, L5–L9 (1997). https://doi.org/10.1093/mnras/289.2.L5
23. P.G. Ferreira, M. Joyce, Structure formation with a selftuning scalar field. Phys. Rev. Lett. **79**, 4740–4743 (1997). https://doi.org/10.1103/PhysRevLett.79.4740. [arXiv:astro-ph/9707286 [astro-ph]]
24. R.R. Caldwell, R. Dave, P.J. Steinhardt, Cosmological imprint of an energy component with general equation of state. Phys. Rev. Lett. **80**, 1582–1585 (1998). https://doi.org/10.1103/PhysRevLett.80.1582. [arXiv:astro-ph/9708069 [astro-ph]]
25. K. Coble, S. Dodelson, J.A. Frieman, Dynamical Lambda models of structure formation. Phys. Rev. D **55**, 1851–1859 (1997). https://doi.org/10.1103/PhysRevD.55.1851. [arXiv:astro-ph/9608122 [astro-ph]]
26. M.S. Turner, M.J. White, CDM models with a smooth component. Phys. Rev. D **56**(8), R4439 (1997). https://doi.org/10.1103/PhysRevD.56.R4439. [arXiv:astro-ph/9701138 [astro-ph]]
27. C. Armendariz-Picon, V.F. Mukhanov, P.J. Steinhardt, A dynamical solution to the problem of a small cosmological constant and late time cosmic acceleration. Phys. Rev. Lett. **85**, 4438–4441 (2000). https://doi.org/10.1103/PhysRevLett.85.4438. [arXiv:astro-ph/0004134 [astro-ph]]
28. C. Armendariz-Picon, V.F. Mukhanov, P.J. Steinhardt, Essentials of k essence. Phys. Rev. D **63**, 103510 (2001). https://doi.org/10.1103/PhysRevD.63.103510. [arXiv:astro-ph/0006373 [astro-ph]]
29. F. Zwicky, Die Rotverschiebung von extragalaktischen Nebeln. Helv. Phys. Acta **6**, 110–127 (1933). https://doi.org/10.1007/s10714-008-0707-4
30. G. Bertone, D. Hooper, History of dark matter. Rev. Mod. Phys. **90**(4), 045002 (2018). https://doi.org/10.1103/RevModPhys.90.045002. [arXiv:1605.04909 [astro-ph.CO]]
31. V.C. Rubin, W.K. Ford, Jr., Rotation of the andromeda nebula from a spectroscopic survey of emission regions. Astrophys. J. **159**, 379–403 (1970). https://doi.org/10.1086/150317
32. L.D. Landau, E.M. Lifshitz, *Mechanics* (Butterworth-Heinemann, Oxford, 1976)
33. V. Demozzi, V. Mukhanov, H. Rubinstein, Magnetic fields from inflation? J. Cosmol. Astroparticle Phys. **08**, 025 (2009). [arXiv:0907.1030 [astro-ph.CO]]
34. A. Proca, Sur la theorie ondulatoire des electrons positifs et negatifs. J. Phys. Radium **7**, 347–353 (1936)
35. P.W. Higgs, Broken symmetries, massless particles and gauge fields. Phys. Lett. **12**, 132–133 (1964)
36. P.W. Higgs, Broken symmetries and the masses of gauge bosons. Phys. Rev. Lett. **13**, 508–509 (1964)
37. F. Englert, R. Brout, Broken symmetry and the mass of gauge vector mesons. Phys. Rev. Lett. **13**, 321–323 (1964)
38. V. Mukhanov, *Physical Foundations of Cosmology* (Cambridge University Press, Cambridge, 2005)
39. G. 't Hooft, Renormalization of massless Yang-Mills fields. Nucl. Phys. B **33**, 173 (1971)

40. G. 't Hooft, Renormalizable lagrangians for massive Yang-Mills fields. Nucl. Phys. B **35**, 167–188 (1971)
41. M.E. Peskin, D.V. Schroeder, *An Introduction to Quantum Field Theory* (Addison-Wesley, Boston, 1995)
42. V. Mukhanov, S. Winitzki, *Introduction to Quantum Effects in Gravity* (Cambridge University Press, Cambridge, 2007)
43. S.L. Glashow, Partial symmetries of weak interactions. Nucl. Phys. **22**, 579–588 (1961)
44. E. Schrödinger, L. Bass, Must the photon mass be zero? Proc. Roy. Soc. (London) **A232**, 1 (1955)
45. D.G. Boulware, W. Gilbert, Connection between gauge invariance and mass. Phys. Rev. **126**, 1563–1567 (1962)
46. A. Komar, A. Salam, Renormalization problem for vector meson theories. Nucl. Phys. **21**, 624–630 (1960)
47. H. Umezawa, S. Kamefuchi, Equivalence theorems and renormalization problem in vector field theory (The Yang–Mills field with non-vanishing masses) Nucl. Phys. **23**, 399–429 (1961)
48. P.A. Ionides, General equivalence theorem in vector field theory. Nucl. Phys. **28**, 662–664 (1961)
49. A. Salam, Renormalizability of gauge theories. Phys. Rev. **127**, 331–334 (1962)
50. M.J.G. Veltman, Perturbation theory of massive Yang-Mills fields. Nucl. Phys. B **7**, 637–650 (1968)
51. D.G. Boulware, Renormalizeability of massive non-Abelian gauge fields: a functional integral approach. Ann. Phys. **56**, 140–171 (1970)
52. A.A. Slavnov, Massive gauge fields. Teor. Mat. Fiz. **10**, 305–328 (1972)
53. M.J.G. Veltman, Generalized ward identities and Yang-Mills fields. Nucl. Phys. B **21**, 288–302 (1970)
54. J. Reiff, M.J.G. Veltman, Massive Yang-Mills fields. Nucl. Phys. B **13**, 545–564 (1969)
55. J.S. Bell, High-energy behaviour of tree diagrams in gauge theories. Nucl. Phys. B **60**, 427–436 (1973)
56. C.H. Llewellyn Smith, High-energy behavior and gauge symmetry. Phys. Lett. B **46**, 233–236 (1973)
57. J.M. Cornwall, D.N. Levin, G. Tiktopoulos, Derivation of gauge invariance from high-energy unitarity bounds on the S matrix. Phys. Rev. D **10**, 1145 (1974). [erratum: Phys. Rev. D **11** (1975), 972]
58. A.I. Vainshtein, I.B. Khriplovich, On the zero-mass limit and renormalizability in the theory of massive Yang-Mills field. Yad. Fiz. **13**, 198–211 (1971)
59. H. van Dam, M.J.G. Veltman, Massive and massless Yang-Mills and gravitational fields. Nucl. Phys. B **22**, 397–411 (1970)
60. M. Fierz, W. Pauli, On relativistic wave equations for particles of arbitrary spin in an electromagnetic field. Proc. Roy. Soc. Lond. A **173**, 211–232 (1939)
61. V.I. Zakharov, Linearized gravitation theory and the graviton mass. JETP Lett. **12**, 312 (1970)
62. Y. Iwasaki, Consistency condition for propagators. Phys. Rev. D **2**, 2255–2256 (1970)
63. A.I. Vainshtein, To the problem of nonvanishing gravitation mass. Phys. Lett. B **39**, 393–394 (1972)
64. C. Deffayet, G.R. Dvali, G. Gabadadze, A.I. Vainshtein, Nonperturbative continuity in graviton mass versus perturbative discontinuity. Phys. Rev. D **65**, 044026 (2002). [arXiv:hep-th/0106001]
65. A. Gruzinov, On the graviton mass. New Astron. **10**, 311–314 (2005). [arXiv:astro-ph/0112246 [astro-ph]]
66. A. Hell, The strong couplings of massive Yang-Mills theory. J. High Energy Phys. **03**, 167 (2022). [arXiv:2111.00017 [hep-th]]
67. A.H. Chamseddine, V. Mukhanov, Mimetic massive gravity: beyond linear approximation. J. High Energy Phys. **06**, 062 (2018). [arXiv:1805.06598 [hep-th]]

68. S.E. Hjelmeland, U. Lindstrom, Duality for the nonspecialist. (1997). [arXiv:hep-th/9705122 [hep-th]]
69. S.R. Coleman, The quantum sine-gordon equation as the massive thirring model. Phys. Rev. D **11**, 2088 (1975)
70. P.A.M. Dirac, Quantised singularities in the electromagnetic field. Proc. Roy. Soc. Lond. A **133**(821), 60–72 (1931)
71. G. 't Hooft, Magnetic monopoles in unified gauge theories. Nucl. Phys. B **79**, 276–284 (1974)
72. A.M. Polyakov, Particle spectrum in the quantum field theory. JETP Lett. **20**, 194–195 (1974)
73. C. Montonen, D.I. Olive, Magnetic monopoles as gauge particles? Phys. Lett. B **72**, 117–120 (1977)
74. E. Witten, D.I. Olive, Supersymmetry algebras that include topological charges. Phys. Lett. B **78**, 97–101 (1978)
75. A. Sen, Dyon - monopole bound states, selfdual harmonic forms on the multi - monopole moduli space, and SL(2,Z) invariance in string theory. Phys. Lett. B **329**, 217–221 (1994)
76. C. Vafa, E. Witten, A strong coupling test of S duality. Nucl. Phys. B **431**, 3–77 (1994)
77. J. Polchinski, Dualities of fields and strings. Stud. Hist. Philos. Sci. B **59**, 6–20 (2017)
78. H. Kawai, A dual transformation of the Nielsen-Olesen model. Prog. Theor. Phys. **65**, 351 (1981)
79. F. Quevedo, C.A. Trugenberger, Phases of antisymmetric tensor field theories. Nucl. Phys. B **501**, 143–172 (1997). [arXiv:hep-th/9604196]
80. F. Quevedo, Duality and global symmetries. Nucl. Phys. B Proc. Suppl. **61**, 23–41 (1998). [arXiv:hep-th/9706210]
81. A. Smailagic, E. Spallucci, The dual phases of massless/massive Kalb-Ramond fields: letter to the editor J. Phys. A **34**, L435–L440 (2001). [arXiv:hep-th/0106173]
82. H. Casini, R. Montemayor, L.F. Urrutia, Duality for symmetric second rank tensors: the massive case. Phys. Rev. D **66**, 085018 (2002). [arXiv:hep-th/0206129]
83. R. D'Auria, S. Ferrara, Dyonic masses from conformal field strengths in D even dimensions. Phys. Lett. B **606**, 211–217 (2005). [arXiv:hep-th/0410051]
84. I.L. Buchbinder, E.N. Kirillova, N.G. Pletnev, Quantum equivalence of massive antisymmetric tensor field models in curved space. Phys. Rev. D **78**, 084024 (2008). [arXiv:0806.3505]
85. D. Dalmazi, R.C. Santos, Spin-1 duality in D-dimensions. Phys. Rev. D **84**, 045027 (2011). [arXiv:1105.4364]
86. M. Shifman, A. Yung, Hadrons of $\mathcal{N} = 2$ supersymmetric QCD in four dimensions from little string theory. Phys. Rev. D **98**(8), 085013 (2018). [arXiv:1805.10989]
87. G.B. De Gracia, Spin jumping in the context of a QCD effective model. Int. J. Mod. Phys. A **32**(06n07), 1750041 (2017). [arXiv:1703.05352]
88. S.M. Kuzenko, K. Turner, Effective actions for dual massive (super) p-forms. J. High Energy Phys. **01**, 040 (2021). [arXiv:2009.08263]
89. J. Louis, A. Micu, Type 2 theories compactified on Calabi-Yau threefolds in the presence of background fluxes. Nucl. Phys. B **635**, 395–431 (2002). [arXiv:hep-th/0202168]
90. Y.M. Zinoviev, On dual formulations of massive tensor fields. J. High Energy Phys. **10**, 075 (2005). [arXiv:hep-th/0504081]
91. M. Gunaydin, S. McReynolds, M. Zagermann, The R-map and the coupling of N=2 tensor multiplets in 5 and 4 dimensions. J. High Energy Phys. **01**, 168 (2006)
92. A. Aurilia, P. Gaete, J.A. Helayël-Neto, E. Spallucci, The bosonized version of the Schwinger model in four dimensions: a blueprint for confinement? Europhys. Lett. **117**(6), 61001 (2017). [arXiv:1702.07294]
93. C. Markou, F.J. Rudolph, A. Schmidt-May, Vierbein interactions with antisymmetric components. J. Phys. Commun. **3**(7), 075013 (2019). [arXiv:1812.07580]
94. C. Markou, F.J. Rudolph, A. Schmidt-May, Gravity with antisymmetric components. Class. Quant. Grav. **36**(9), 095014 (2019). [arXiv:1811.12419]
95. L. Heisenberg, G. Trenkler, Generalization of the 2-form interactions. J. Cosmol. Astroparticle Phys. **05**, 019 (2020). [arXiv:1908.09328]

96. A. Smailagic, E. Spallucci, Cornell potential in Kalb-Ramond scalar QED via Higgs mechanism. Phys. Lett. B **803**, 135304 (2020). [arXiv:2002.12750]
97. A. Hell, On the duality of massive Kalb-Ramond and Proca fields. J. Cosmol. Astroparticle Phys. **01**(01), 056 (2022). [arXiv:2109.05030 [hep-th]]
98. C.N. Yang, R.L. Mills, Conservation of isotopic spin and isotopic gauge invariance. Phys. Rev. **96**, 191 (1954)
99. S. Weinberg, The making of the standard model. Eur. Phys. J. C **34**, 5–13 (2004). [arXiv:hep-ph/0401010 [hep-ph]]
100. M.D. Schwartz, *Quantum Field Theory and the Standard Model* (Cambridge University Press, Cambridge, 2014)
101. A. Accioly, J. Helayel-Neto, E. Scatena, Upper bounds on the photon mass. Phys. Rev. D **82**, 065026 (2010). [arXiv:1012.2717 [hep-th]]
102. W. Greiner, J. Reinhardt, *Field Quantization* (Springer, Berlin, 1996)
103. P.A.M. Dirac *Lectures on Quantum Mechanics*. Belfer Graduate School Monograph Series, no. 2 (Yeshiva University, New York, 1964)
104. S. Deser, Ann Inst. Henri Poincare, Vol.XVI, Paris, Gauthier-Villors, pp. 79 (1972)
105. L. Heisenberg, J. Zosso, Quantum stability of generalized proca theories. Class. Quant. Grav. **38**(6), 065001 (2021). [arXiv:2005.01639 [hep-th]]
106. M. Veltman, *Diagrammatica: The Path to Feynman Diagrams*. Cambridge Lecture Notes in Physics (Cambridge University Press, Cambridge, 1994)
107. R.E. Cutkosky, Singularities and discontinuities of Feynman amplitudes. J. Math. Phys. **1**, 429–433 (1960)
108. M.J.G. Veltman, Unitarity and causality in a renormalizable field theory with unstable particles. Physica **29**, 186–207 (1963)
109. E.C.G. Stueckelberg, Die Wechselwirkungskräfte in der Elektrodynamik und in der Feldtheorie der Kräfte. Helv. Phys. Acta **11**, 225 (1938)
110. T. Kunimasa, T. Goto, Generalization of the stueckelberg formalism to the massive Yang-Mills field. Prog. Theor. Phys. **37**, 452–464 (1967)
111. P.T. Matthews, The S-matrix for Meson-Nucleon interactions. Phys. Rev. **76**(8), 1254–1255 (1949)
112. J. Gegelia, G. Japaridze, On renormalizability of the effective field theory of massive Yang-Mills fields. Mod. Phys. Lett. A **27**, 1250128 (2012). [arXiv:1109.3880 [hep-th]]
113. S. Wong, Massless limit of the massive Yang-Mills field. Phys. Rev. D **3**, 945–952 (1971). [erratum: Phys. Rev. D **3** (1971), 3243–3243]
114. A.A. Slavnov, L.D. Faddeev, Massless and massive Yang-Mills fields. Theor. Math. Phys. **3**, 312–316 (1970)
115. D.G. Boulware, S. Deser, Can gravitation have a finite range? Phys. Rev. D **6**, 3368–3382 (1972)
116. R. Ferrari, A. Quadri, Physical unitarity for massive non-Abelian gauge theories in the Landau gauge: stueckelberg and higgs. J. High Energy Phys. **11**, 019 (2004). [arXiv:hep-th/0408168 [hep-th]]
117. D. Bettinelli, R. Ferrari, A. Quadri, A massive Yang-Mills theory based on the nonlinearly realized gauge group. Phys. Rev. D **77**, 045021 (2008). [arXiv:0705.2339 [hep-th]]
118. D. Bettinelli, R. Ferrari, A. Quadri, One-loop self-energy and counterterms in a massive Yang-Mills theory based on the nonlinearly realized gauge group. Phys. Rev. D **77**, 105012 (2008). [erratum: Phys. Rev. D **85** (2012), 129901] [arXiv:0709.0644 [hep-th]].
119. N. Dragon, T. Hurth, P. van Nieuwenhuizen, Polynomial form of the Stuckelberg model. Nucl. Phys. B Proc. Suppl. **56**, 318–321 (1997)
120. H. Ruegg, M. Ruiz-Altaba, The Stueckelberg field. Nucl. Phys. B Proc. Suppl. **56**, 318–321 (1997). [arXiv:hep-th/9703017 [hep-th]]
121. L. Alberte, A.H. Chamseddine, V. Mukhanov, Massive gravity: resolving the puzzles. J. High Energy Phys. **12**, 023 (2010). [arXiv:1008.5132 [hep-th]]
122. V.A. Rubakov, P.G. Tinyakov, Infrared-modified gravities and massive gravitons. Phys. Usp. **51**, 759–792 (2008). [arXiv:0802.4379 [hep-th]]

123. V.F. Mukhanov, H.A. Feldman, R.H. Brandenberger, Phys. Rep. **215**, 203 (1992)
124. K. Hinterbichler, Theoretical aspects of massive gravity. Rev. Mod. Phys. **84**, 671–710 (2012). [arXiv:1105.3735 [hep-th]]
125. G.R. Dvali, G. Gabadadze, M. Porrati, 4-D gravity on a brane in 5-D Minkowski space. Phys. Lett. B **485**, 208–214 (2000). [arXiv:hep-th/0005016 [hep-th]]
126. G. Dvali, A. Gruzinov, M. Zaldarriaga, The accelerated universe and the moon. Phys. Rev. D **68**, 024012 (2003). [arXiv:hep-ph/0212069 [hep-ph]]
127. C. de Rham, Massive gravity. Living Rev. Rel. **17**, 7 (2014). [arXiv:1401.4173 [hep-th]]
128. I.B. Khriplovich, Green's functions in theories with non-abelian gauge group. Sov. J. Nucl. Phys. **10**, 235–242 (1969)
129. A.H. Chamseddine, V. Mukhanov, Massive hermitian gravity. J. High Energy Phys. **08**, 036 (2012). [arXiv:1205.5828]
130. T.L. Curtright, P.G.O. Freund, Massive dual fields. Nucl. Phys. B **172**, 413–424 (1980)
131. T.L. Curtright, Massive dual spinless fields revisited. Nucl. Phys. B **948**, 114784 (2019). [arXiv:1907.11530]
132. T.J. Allen, M.J. Bowick, A. Lahiri, Topological mass generation in (3+1)-dimensions. Mod. Phys. Lett. A **6**, 559–572 (1991)
133. E. Harikumar, M. Sivakumar, Duality and massive gauge invariant theories, Phys. Rev. D **57**, 3794 (1998) [arXiv:hep-th/9604181]
134. A. Smailagic, E. Spallucci, Duality of massive gauge invariant theories in arbitrary space-time dimension. Phys. Rev. D **61**, 067701 (2000). [arXiv:hep-th/9911089]
135. G. Dvali, Three-form gauging of axion symmetries and gravity (2005). [arXiv:hep-th/0507215]
136. M. Kalb, P. Ramond, Classical direct interstring action. Phys. Rev. D **9**, 2273–2284 (1974)
137. R.L. Davis, E.P.S. Shellard, Antisymmetric tensors and spontaneous symmetry breaking. Phys. Lett. B **214**, 219–222 (1988)
138. G. Dvali, S. Hofmann, J. Khoury, Degravitation of the cosmological constant and graviton width. Phys. Rev. D **76**, 084006 (2007). [arXiv:hep-th/0703027 [hep-th]]
139. A. De Felice, L. Heisenberg, R. Kase, S. Tsujikawa, Y. l. Zhang, G. B. Zhao, Screening fifth forces in generalized Proca theories. Phys. Rev. D **93**(10), 104016 (2016). [arXiv:1602.00371 [gr-qc]].
140. S. Ansoldi et al., Dualization of interacting theories including p = d - 1 limiting cases. Phys. Lett. B **471**, 133–139 (1999). [arXiv:hep-th/9911051]

The manufacturer's authorised representative in the EU is Springer
Nature Customer Service Centre GmbH, Europaplatz 3, 69115 Heidelberg,
Germany. If you have any concerns regarding our products, please
contact ProductSafety@springernature.com

Printed and bound by CPI Group (UK) Ltd, Croydon, CR0 4YY
01/10/2026
02225982-0008